W. Livingston Harlan

Osteopathy, the new science

W. Livingston Harlan

Osteopathy, the new science

ISBN/EAN: 9783337712051

Printed in Europe, USA, Canada, Australia, Japan

Cover: Foto ©berggeist007 / pixelio.de

More available books at **www.hansebooks.com**

OSTEOPATHY;

THE NEW SCIENCE.

———

By W. LIVINGSTON HARLAN, D. O.,

PRESIDENT OF THE

ROCKY MOUNTAIN INFIRMARY,

DENVER, COLORADO.

———

CHICAGO:
1898.

INTRODUCTION.

This little volume, which we believe is the first that has been published in book-form, on the science of Osteopathy, was written with a two-fold purpose. First: To inspire a greater love for Osteopathy in those who have taken it up as a life-work. Second: To give those who are not acquainted with the new science some idea of its teachings, and something of its importance to suffering humanity.

In compiling it, we have quaffed generously from that fountain of knowledge—"The Journal of Osteopathy," whose proprietor and publisher, Dr. Andrew Taylor Still, is the discoverer of the science, and who, as a thinker and worker in the new field, stands unrivaled.

At the close of this the 19th century, when modern thought is doing so much for civilization, Osteopathy presents its claims for serious consideration, feeling confident that an enlightened public will give it a befitting welcome, as the art of healing without drugs.

W. L. H.

Denver, Feb., 1898.

"He that hath truth on his side is a fool as well as a coward, if he is afraid to own it."

—De Foe.

W. Livingston Harlan

OSTEOPATHY, THE NEW SCIENCE.

DR. W. LIVINGSTON HARLAN.

By FRANK TREBBLE.

Although Dr. Harlan has been in Denver only about two years, he has met with phenomenal success as an Osteopathic physician. His success and popularity are due to his superior skill in his profession, coupled with a conscientious desire to do his utmost to help suffering humanity.

He was born at Kirksville, Missouri, February 26, 1872. He is the son of W. B. Harlan, a retired merchant of that city. After attending the public school and also the Normal school, he was graduated from the commercial college at Kirksville in 1889. He then traveled two years for a prominent medical and electrical company; then engaged in the mercantile business in his native town, and after occupying the position of express agent for a short time at the same place, he embarked in the mining business at Webb City, Missouri, where he owned and operated the lead and zinc mines on McCorkle Hill. Returning home, he matriculated at the American School of Osteopathy, where he took a three years' course, receiving his diploma in 1896. He then located in St. Louis, remaining there about six months. While there he treated many of the best-known and wealthiest families of that city. He came to Denver the same year, where he soon organized the Rocky Mountain Infirmary Association, of which he is the active principal.

7

Seeing the necessity for educating the people in the new science, he started "The Rocky Mountain Osteopath," a monthly and weekly paper, and the first paper of the kind published in the state, and the first weekly that was published in the world.

While a mere youth, the doctor developed a fondness for the study of anatomy, and a desire to understand the anatomical construction of the human frame.

The turning lathe, and all sorts of machinery and industrial inventions were always of great interest to him. Although he served no apprenticeship in any of the mechanical arts, his genius was such that he could readily construct anything of that nature to which he turned his attention.

He has sat at the feet of that modern Gamaliel, Dr. A. T. Still, the founder of the American School of Osteopathy, nearly all his life. Being thoroughly familiar with Dr. Still's professional methods in treating patients, he has had advantages in studying and perfecting himself for the role he has assumed, that rarely fall to the lot of any young man in professional life. He is the only Osteopathic physician who was born and raised at Kirksville (the home of Dr. Still) who is now following his profession in the West.

The doctor is not only a natural anatomist, the study of which forms the basic principle of Osteopathy, but he is devoted to his profession, and being a man of energy and determination of purpose, he is rapidly ascending the professional ladder to fame and fortune.

He is both aggressive and tenacious of his convictions, and lets no opportunity pass, where he can do or say something that will tend to the legitimate good of his profession.

Others may claim the distinction of possessing diplomas

from the Kirksville School, but Dr. Harlan is the only post graduate of that institution who is now practicing in the West. He is an able and earnest representative of the school of Osteopathy, and has done much to inculcate its teachings among the cultured and intelligent people of Denver.

Although quite young (not yet twenty-seven years of age) the doctor has followed a variety of pursuits, and inaugurated a number of paying enterprises, but none of these satisfied him. He never found anything that suited him, or anything that he could willingly adopt as a life-work, until he began the study of Osteopathy. This was found to be in complete harmony with the ideas that he had been imbibing from his youth up, and the more he read of anatomy, and the longer he attended the American School of Osteopathy, the more deeply he became impressed with the conviction, that his present profession was the one that he should choose.

Some men are especially endowed with certain qualifications for the pursuits they follow. The doctor is one of these. He possesses an elastic, vigorous physique, which is capable of undergoing almost any fatigue. He is a close and keen observer of men, and possesses most excellent judgment in all the practical affairs of life, and the fact that he has been financially successful, shows him to be a man of good business capacity.

Like thousands of the most active and prominent men of our times, the doctor when beginning life for himself could boast of none of this world's goods. Possessing, however, two strong arms and a brave heart, he always worked with a determination to win, and success has crowned his efforts. He is easily approached, is a pleasant conversationalist, and handles his patients so deftly and skillfully that he at once inspires their confidence.

The suite of rooms occupied by him as offices, are among the largest and most handsomely fitted up to be found in our city, and being fond of literature, his tables contain the latest magazines and papers of the day. He is methodical and painstaking, and keeps a complete record of not only his business transactions, but a record of all his patients, of their treatment, the day and hour they were treated and the disease from which they were suffering.

OSTEOPATHY IN THE COLORADO LEGISLATURE.

The fact that Osteopathy is recognized in Dakota, Missouri, Michigan and Vermont, shows that it is gradually winning its way before an enlightened public. The bill in reference to Osteopathy, passed the lower House of the Missouri Legislature 106 to 16 votes, and the Senate 26 to 3 votes, and immediately received the signature of Governor Stevens of that state. The bill passed the legislatures of the different states by a large majority.

The now famous bill, which was introduced into the Colorado Legislature of 1896-7, was framed by Dr. W. Livingston Harlan, of Denver. Notwithstanding the bitter antagonism which was exhibited by the members of the medical profession, the bill received in the lower house, 41 votes, there being 11 against it, and in the senate, 22 votes to 9 against it. It was, however, vetoed by Governor Adams on the ground that it was too new a science to be thus recognized. The governor said by implication, that anything that was new possessed no merit. The governor, we doubt not, would decline to wear a second-hand suit of clothes, that some tramp had used, but in the case of the Osteopathic bill, he, perhaps, thought that the science of Osteopathy should be worn out by fakes, quacks, and charlatans, before its genuine and devoted followers should be permitted to practice it. His veto was interposed without any reasonable cause, and the grounds of his opposition partook more of the nature of puerile objections than of sound argument. The honest, intelligent people of advanced thought could hardly understand why the science of Osteopathy, which had been endorsed by the general assemblies of four different states, should

meet with any opposition from our supposed to be modern,
up-to-date chief executive, especially after it had passed
both houses of our legislature by such an overwhelming
majority. The presumption is, that the legislators knew
what they were doing, Governor Adams to the contrary,
notwithstanding. All honor to the members of the general assembly of Colorado, who had the courage of their
convictions, and voted for the bill!

The purposes of the Osteopathic bill as intended by its
author, were, to create a state examining board, which
should govern in all things pertaining to the practice of
Osteopathy in Colorado, instead of permitting the enemies
of the science to pass upon the technical, or other qualifications that might be required by law. The bill also provided that all practitioners of the science should be compelled to obtain a thorough knowledge of the same, by passing through a course of study, and securing a diploma
preparatory to their entering upon the practice. The bill
was wholly in unison with the requirements and teachings of the American School of Osteopathy, of which Dr.
A. T. Still is the founder, and who is also the discoverer
of the science. The American is the best school of Osteopathy in existence, and is regarded by the Rocky Mountain
Infirmary of Denver as such. It has the advantage over
all other schools, from the fact that its head and man-
ager, Dr. Still, is recognized as the ablest man in the
practice of Osteopathy. His many years of experience as
an old line physician, his strong, vigorous intellect, and
natural endowments, pre-eminently fit him for the high
position he occupies.

It is proper that all bills which concern Osteopathy
should represent the discoverer's ideas; especially when
it has been shown by experience that his ideas are in full
accord with the teachings of the new science, and that

his mode of treatment of disease has proven to be the most successful now extant. Notwithstanding the apparent.unselfish purposes of the Colorado bill, we regret to say that a few Osteopaths, some of whom were graduates of the school, most bitterly opposed it as it was originally framed by its author, and finding they could not successfully defeat it in this way, tried in various ways to so amend its provisions as to finally defeat its legitimate intent. It is taken for granted by all well educated Osteopathic physicians, and intelligent people generally, that persons having no diploma from some reputable school of Osteopathy, and are practicing, that they are generally mere fakes, or charlatans, and the conclusion is inevitably drawn, that they are imposing themselves upon the credulous public. That numbers of people in Colorado are being subjected to this species of humbuggery, there can be no doubt, and if these pretenders were called upon to exhibit their diplomas from some recognized school of Osteopathy to their dupes, a less number of innocent people would be bled out of their hard earnings. There would be less of mal-practice, less number of bones broken, or dislocated in the human frame, and less of pain and suffering as the result.

The bill mentioned, was to prevent mal-practice by pretenders, and to eliminate from the profession the well-known shysters. In the face of all these facts, Governor Adams saw fit to veto the bill, and thereby gave free rein to the Osteopathic quacks, who are by their methods, bringing this new and wonderful science into disrepute.

It is amusing to read Senator Reuter's speech in opposition to the bill. His was the chief argument, if we may dignify it as such, that was made against it.

His speech, which was delivered in the senate, March 19, 1897, will be here reproduced in full, that our readers may get an intelligent idea of the reasons why he opposed it:

"Mr. President: This is one of the bills that is passed under the extraordinary circumstances of a rush; one of these moods we get into occasionally; we get together tired and weary, often late at night, and a measure of this kind is sprung on us, and it goes through without proper consideration.

"I haven't very much to say about this bill in addition to what I stated the other evening, but I do desire to call attention to some of the peculiar features of this bill. In the first place, this bill speaks of the science of Osteopathy, as if there were in existence such a science. There is no such science as the science of Osteopathy; you will not find the term in any of the dictionaries or encyclopædias, and it is extraordinary that any senator upon this floor should be permitted to have himself persuaded that a thing which is not yet old enough to receive baptism; a thing that you can't hear spoken of; that you can get no literature upon should be entitled to be designated by the term 'science.'

"I understand that the methods pursued by this so-called science is nothing more, is nothing else than a species of massage Swedish movement. Swedish movement is older than the mountains. Swedish movement is discussed in scientific works, and from all that I have heard of Osteopathy, upon this floor or elsewhere; all that I have been able to learn from one of these fancy, handsome quack pamphlets, that were sent throughout the country, in which they claim to define Osteopathy, I find absolutely nothing in it, except, that Osteopathy is something that will cure. They there give a whole page of three columns, of all the various troubles that can be cured by this so-called science.

"They offer the explanation—they ask you to read this

little pamphlet, this little quack advertisement, and from
that you can understand what is the science of Osteopathy.
I had that circular here the other day, but I haven't it here
now; but I think every senator upon this floor has received
from two, to a half dozen of these circulars, but I trust, if
there is a senator here who believes, or who will claim
either that this is a science, or that there is anything in
this circular which explains what it means, I trust he will
rise and give us that explanation.

"From the talk that is made upon the floor here, it seems
that many senators for the first time have heard that the
blood courses through a man's body, and that it courses
through a man's body through veins and arteries; and one
of the prominent gentlemen of this city in a little discussion the other day, called my attention to the fact, that he
had been convinced that this was a science, from a little
book that he had read, and in that little book, when he
called my attention to it, appeared this wonderful discovery, that had never been heard of before, and what was
it? He said the human body can be compared to a court
lawn, and in the center of that lawn you can imagine
there is a lake, and imagine from that lake, pipes running
out onto this lawn that carried water, and so long as the
water courses through these pipes, that lawn will grow,
and when those pipes are stopped up, or any one of them,
that portion of the lawn to which that little pipe leads
must of necessity die. That explanation, Mr. President,
is not bad. There is no doubt in the world but what, in
order to have healthful life in the human body, it is necessary that the blood circulate freely through the body.
There is no question in the world but what, if an artery
were cut or clogged up with something, would break, something would get hurt, a man would not be well.

"The same with the veins than run the blood to the

heart. But, Mr. President, any man who is so ignorant, any man who knows so little about medical science, or about the human body that he has to be told that as an argument, that Osteopathy is a science, I say, ought to hesitate before he undertakes to pass judgment on any science. Read the books upon Swedish movement and science, you will find that the blood must flow freely through the veins and arteries in the body, not only blood, but other liquids that are necessary in the joints and other parts of the body, must be removed if necessary and carried off. You will find a full discussion of this proposition in all books upon the Swedish movement. It is upon that that the Swedish movement is based, and it is upon that that Osteopathy, so far as I can learn, is based. The basis is a correct one; there is no question in the world but what a species of massage is always good for particular cases. I would venture to say, Mr. President, that it would be better for all of us, if used from time to time.

"Physical exercises are part of the same thing. It is simply parts of science; there are books upon books on physical exertion, upon how you shall take it, how you shall move your muscles, move muscles of your hand and body, of your limb, the muscles of your neck and all that sort of thing, in order to produce a free circulation of the blood through the body. To that extent, I have no objection to Osteopathy. I do not say that Osteopathy when properly applied, and applied under proper medical advice, and proper conditions, is not healthful and beneficial. But who ever heard of a doctor of massage? Who ever heard of a Swedish movement seeking to elevate itself to being an entirely distinct and separate science, as being one upon which you could base such a bill as this bill here is based? So that I say to call this a separate science, and call this a separate system of treating the human body is absurd.

"I say if this matter of Osteopathy shall receive any kind of recognition, the only kind of recognition it should receive is that no man shall practice it, excepting under the directions of a physician, or unless he be one who has himself studied and completed a course of medicine in all its branches and courses. Any man who says he can treat any part of the human body, or any one of the organs of the human body, without understanding the entire human body, without understanding the physiology and anatomy of the human body, and the things that are necessary to learn in the science of medicine—I say any man who thinks he can learn one of these little things that are claimed to be contained in Osteopathy, and be qualified to treat the human body, is a quack, is a shyster, is a pettifogger in his profession.

"Let us get to the provisions of this bill for a moment. What does this bill provide? I want to direct the attention of the Senate to this proposition, that this bill is a money-making scheme. It originated in Kirksville, Missouri; it originated so far as I know, with an unknown man. This method was probably developed by that man. He has probably studied more closely than is studied in the schools of medicine to-day this method of treatment, but he is not the only man who is treating with it, and that man, I understand, is a humanitarian of the first order, and he makes no charge to the rich or poor for this treatment. He is a man of education I understand, and who, when people come to him, and he recognizes that it is not within his sphere, and they cannot be treated by him, he turns them back. He does not belong to this class of quacks that go through the country, and try to push a bill of this kind through the legislature. He is not the

2

kind of man who goes forth to get a measure of this kind
into statute in this state—

Sen. Moody: "Who is that?"

Sen. Reuter: "I do not know his name. I have no
fault to find with that gentleman, but when these people
start out on this course; when they leap from legislature
to legislature in this country; when they seek to get something here which would bring money into their pockets;
when they conduct their methods with every species of
advertisements, with every species of deception that is
known to quackery, then I say it is time for us to call a
halt upon an institution of that kind. This bill provides
this extraordinary thing; it appoints one man a commis-
sion to judge, and makes him a board to examine the applicant. Our statute nowhere puts the power in one man
and makes him a board to determine this important thing.
Let us see what he is to determine. He is the sole and
absolute judge as to the qualification of the man that
comes before him. This bill provides that if that man
brings a diploma from a reputable institute; if he comes
from one of these institutions that is said to be reputable
with a diploma, or appears reputable to his individual judgment, he can give that man a diploma; if he does not,
he refuses him a diploma. If he gives him a diploma, he
gets a fee of five dollars. Where are these reputable institutions throughout the country that give these diplomas?
There is nothing in this bill that requires that these men
who bring these diplomas from one of these institutions,
shall have any course of study. It does not provide that a
party who presents such a diploma shall have studied one
day or one hour in that institution; a diploma is all he
needs, and if he gets his diploma, all that the said examiner
needs, is five dollars. We all know how we have sought
to cast around the people some kind of a safe-guard. We

require here an examination of learned men, six or seven men. When a man comes into this state to practice medicine, they shall determine upon the sufficiency of his diploma, and why? Because as medical institutions are constituted throughout the country, and also in other countries, this issuing of diplomas to people who are wholly incapable of practicing medicine was a common thing. They were established for the sole purpose of giving diplomas; they were money-making institutions, and they issued their diplomas to any man who would go there and go through the formality of study and pay them; twenty-five, fifty or one hundred thousand have a diploma of that kind, and these men were turned loose upon the community."

Sen. Moody: "Fortunately this is not a medical school."

Sen. Reuter: "The only good that I can see of this not being a medical school is that a man can get through it without studying anything, without knowing anything at all, and with having but twenty-five dollars to get a diploma."

Sen. Barela: "Do you know how long a man must study to get a diploma in Osteopathy? They study between two and three years as I am told upon respectable authority."

Sen. Reuter: "Where do you find that? The gentleman always cites somebody outside who is as respectable as the gentleman inside."

Sen. Barela: "The reason I say that some man outside may have told you."

Sen. Reuter: "I do not like to bring the names of citizens into a matter of this kind, nor do I think that it is necessary, and so far as I am concerned, I think it is out of the case, but, we may differ on that proposition. I have no desire to quote any gentleman's name. I shall simply quote what seemed to somebody a sufficient reason for calling this thing a science, and had nobody told me, and if I knew now that the senator was correct in his suggestion, but I do not like the use of these names. The senator says that somebody told me that before these particular people that are here now practicing the so-called Osteopathy, they had studied for several years in Missouri, probably at Kirksville.

"We will say these gentlemen who are here now may be persons who studied under this man that I spoke of, and those persons may be competent to carry out this branch of the science; that may be; that is not the argument I am making, but it is that this man here, who may be by chance appointed commissioner, will permit nobody to practice here who has not studied two or three years. If this bill provided that no certificate should be received by that commissioner, unless the school from which that diploma came taught a certain course, taught a certain length of time, there might be some little safeguard in that, but this simply provides that if in his particular judgment, if it is a reputable school, that then he would be authorized to deliver his certificate for the little fee of five dollars.

"If there is a senator upon this floor who believes that the course of study laid down in section 5 is a course of study which will qualify a man to treat the human body, then I have nothing more to say to that senator, excepting this, that it is absurd to me, it is nonsense to me, that a man who can go through the study of these things says he is capable of treating the human body for any disease. If this is a science, no man in favor of it dares to claim that a man can treat any portion of the human body, or any of its functions, without knowing all, because every function of the human body, every vessel in his body, every tissue in his body, is not an independent one, and all of these things are co-dependent and interdependent upon each other, and no man can diagnose a case without know-ing the entire human system in all its features and pos-sibilities.

"That is what medical science is, so that if a man who just practices this—if a man, for instance, coming to one of these adventurers with a certain trouble, how is he to diagnose what that trouble is? and until he has diagnosed it and knows what this trouble is, he cannot intelligently apply his remedy. The first thing that a physician must know when he seeks to treat a patient, is, what is the matter with the patient. The first thing one of these parties practicing this so-called Osteopathy must know, is, what ails that person, and how can he tell correctly; what foundation has he for the belief even of what ails that man, unless it is something that is very apparent upon the surface of the body, or something of that kind? If it is all exterior! This provides here that they shall be taught to treat the body against noxious drugs. How in the world can they tell anything about noxious drugs? or the effect of them without having studied them? The whole course of study laid down here is absurd, an absurdity when you choose to elevate this thing into a science. And what does it say here? It simply says if a school of that kind is established in this city, then that course must be pur-sued. And if a little establishment for the purpose of making money establish itself here beyond the boundaries, in a state adjoining or the territory of New Mexico, abso-lutely no study is required. If you want to draw a bill for the express purpose of giving some one man in the state of Colorado the monopoly of admitting to this prac-tice; if you want to give to one man the right to determine that some particular school in New Mexico or Wyoming, or Missouri, shall have the absolute monopoly of turning out these men, you have given it in this bill, and you have thrown absolutely no safeguards around it, because if he wishes to go to a school in Missouri he can say it is the only reputable school in the country. If he wants to promote that school, and say some other school in Nevada is the only reputable school, that man can promote that school, or any other institution in the country, and what can he do when you place a single man in that position? When you place him there without the power of the governor to remove him, you place him there absolutely for two years.

"Mr. President, the very fact that the party who desires the situation, as I understand, was upon the floor of this senate and insulted members of this body, because of the reduction of the six years down to two years furnishes no better reason to object to this bill than that it is intended for an absolute monopoly. Let us go a little further. If I am correct in the proposition that this is not an intelligent science; if I am correct in the proposition that no man should be permitted to practice these movements, and saying a man shall be treated by that method unless he is a physician, then I say it is base quackery, and what do you say? I call attention again to the modes of advertisement of this suspicious work. You will find it in newspapers; you will find it in flaming advertisements up over the building near here; you will find it in expensively printed circulars that they send over the country; you will find in that circular that there is not an evil that the flesh of man is heir to, that it don't claim to be able to cure by that kind of treatment. And what is the result if we stamp upon that the seal of the state of Colorado and say that the house of representatives and senate of Colorado have endorsed this scheme?

"I have no fear of the success of this scheme of medicine upon the intelligent and richer classes of the state. What

I object to in this law is for the same reason as I do to all other laws of the same kind; they are to protect the weak who cannot by their own investigations know the difference between quackery and medicine. When you place the stamp of the state upon this law, the next circular that goes forth will devote the entire title page on the front to the fact and call the attention of the poor people to it, that this has been recognized by the state of Colorado, and in that way they will extort their fees from people who are unable to pay them—people who will go there with the hope of being treated, and say if the state of Colorado has endorsed this scheme it must be a good one.

"I do not know the gentlemen who are here. I am not personal in my remarks in regard to the particular gentlemen who are here. I have no cause for any enmity towards these people. I am speaking generally and upon the proposition of what human is, and what is likely to be done under circumstances of this kind based upon the experience of years, that it must inevitably follow such an act as this as night follows day; that the people will be imposed upon by this kind of quackery. Quackery I use only in that sense when this particular method is applied by persons who generally do apply it, and when it is applied for purposes for which it should not be applied. I think I have made myself understood that there can be absolutely no question about it, but that treatment of that kind under certain circumstances and properly directed is beneficial, and so far as medical men are concerned that is not denied. You will find chapter after chapter upon these same propositions, but until this thing has been sought to be exalted into a money-making scheme nobody ever dreamed of calling this thing a separate and independent science.

"Mr. President, in addition to that I desire to call attention to this further feature. Up to this time massage is constantly applied; massage is constantly prescribed by the medical profession. There are physicians who, when they have no skill in massage will refer you to an operator who has, but what does this bill do? This bill provides that notwithstanding a man may have graduated in the science of medicine, and practiced medicine for twenty-five or fifty years, and may be a man at the head of his profession, under this bill he is prevented from applying a method which he may have applied from the time that he was first admitted to practice medicine without going before this commissioner or examiner of the state, a man who is not required to be a physician; a man who is not required to be a scientific man; a man who is not required to have done anything more than for two years to have studied this little stuff that is provided for in the statute; a man who may not have even studied this stuff prescribed, because the bill does not even say and provide that the examiner shall be a man who has studied it, and I say that if a physician cannot by any possibility under this law practice something that has been known to medicine for years and years without going before this examiner and submitting himself to an examination, who, when he does go before him and submit himself to an examination this examiner will say to him: 'Where is your certificate from the school of Osteopathy under which I can admit you to do that which you have done for years and years?' So you will have to go to some school of Osteopathy for two years before you will be permitted to treat your patient in a method in which all scientific books on medicine have taught for centuries.

"I say, Mr. President, that the passage of such a bill as this will be a stigma upon our good name. Senators can tell you, I believe, that a similar bill passed in Missouri.

It may have been passed in Missouri. There are lots of little theories and quackeries that every now and then come up which simply seem to appeal to the sentiment of most people, and I say if we consider this matter seriously, if this were a provision that no man should practice Osteopathy without having gone before our medical examiners of the state of Colorado, and having been examined upon even the studies that are prescribed here, and that that board of medical examiners find that the party who applies for the purpose of treating in Osteopathy has had sufficient experience in the operation of that peculiar method that he is entitled to practice there would be some sense in it, but in this bill, in my opinion, it would be a disgrace to the state, and for that reason I shall vote against the bill."

It will be observed that the senator begins his speech by saying that he hadn't much to say about the bill and then proceeds to make a long, labored argument which perhaps covers the space of more than an hour for the purpose of convincing senators that the bill he attacks is a disgrace to the state.

He facetiously remarks that the bill "speaks of the science of Osteopathy, as if such a science were in existence." The senator shows his ignorance of the subject, for Osteopathy is not only a science but had been recognized as such by four different states of the Union, and had at that time been taught by a number of schools and colleges, one of which had been in existence no less than ten years. These institutions were founded and maintained for no other purpose than that of preparing students by a rigid and thorough training for the practice of this very science. He then discloses the fact which was doubtless wonderful to himself, that there was no such term in any of the dictionaries or encyclopædias as Osteopathy,

and therefore thought it most extraordinary that any
(Colorado) senator should be persuaded that there was
such a science. The same method of reasoning might
have convinced Columbus that there was no new world
because he could not find any on the maps.

The senator says that this so-called science is nothing
else than a species of Swedish movement, and then admits
that the Swedish movement is as old as the mountains.
Continuing, he says that the Swedish movement teaches
that the blood and other liquids of the body must flow
freely through the veins and arteries, and that so far as
he could learn osteopathy is based upon the same theory,
and says that the basis is a correct one. In other words,
his argument is this: The Swedish movement is all right;
Osteopathy is based upon the Swedish movement, but
Osteopathy is all wrong. His premise is well laid, but he
deduces false reasonings therefrom. Osteopathy is neither
the Swedish movement nor can it be classed as massage,
for the latter can be performed by anyone who has not the
remotest idea of the science of anatomy, which is the basic
principle of osteopathy. The most unskilled, the most
unscientific person may use the massage treatment, which
is done by rubbing. Osteopathy upon the other hand,
implies if the doctor administering it (is not a quack) a
thorough knowledge of anatomy, which requires years of
study with practical demonstrations in one of the schools
which teaches that science exclusively.

The thorough trained Osteopath is not only familiar with
anatomy, but he understands the construction of the
human frame in all its minutia; the exact location and
conformation of the bones; their sizes, texture, terminal
points; the relation they sustain to each other; the veins,
the arteries, the flow of blood, the normal and abnormal
conditions of every part and particle of that most wonder-

ful of all machines, the human body. His diagnosis is the result of both the use of his visual organs, and his hands, and more especially the latter, as the cause of disease or pain lies generally under the surface where it cannot be seen by the eyes. The bare touch of the skilled Osteopath will in most cases, locate in an instant the spot whence comes the pain and enables him to intelligently treat the patient by his knowledge of anatomy and physiology. It comes with bad grace from a non-professional man; one who is totally ignorant of the subject he attempts to discuss (Osteopathy) to characterize those who practice it as quacks and shysters. But all the great leaders of thought and their theories were at first denounced as visionaries and ignoramuses, simply because they were in advance of their day and time. The senator, however, betrays his own ignorance in a wonderful manner when he admits that he knows nothing about a science which was discovered nearly twenty-eight years ago, and which has since that time received practical embodiment in a number of permanently endowed schools and colleges.

The Swedish movement was nothing more than the modern massage, which has been revamped to suit modern taste and credulity. Had Senator Reuter extended his researches a little further he might have found the term or word "osteopathy" in Gould's medical dictionary, and had he examined that greatest of all lexicons, published by Funk and Wagnalls, he would have found the word "osteo" derived from the Greek word osteon, which means a bone. Admitting, however, that this word cannot be found in the dictionaries this proves nothing for thousands of new words have been coined during the past two decades, words which are entirely new even to the senator's vocabulary.

The senator is again wide of the mark when he intimates that the fees of the Osteopath will be "extorted" from poor people. The most intelligent as well as the wealthiest class of people in Colorado, as is the case elsewhere, are largely in the majority among those who are the patients of the Osteopathic doctors. These patients are neither poor nor ignorant, but are people of culture and refinement and are generous enough to believe that good may even be found in Osteopathy. They have tried it and are loud in their praises of what this science has done for them. Had the senator interviewed a few of these patients before he made his speech, which is composed largely of guesses, he, perhaps, might have spoken more to the point and had a more intelligent understanding of the theme he attempted to discuss.

The senator says in substance that any man who can treat any part of the human body or any one of the organs of the human body without understanding the anatomy and physiology of the human body, etc., is a quack, shyster and pettifogger. We heartily agree with him, and say that such a pretender should be denounced as a quack. But the Osteopath, that is, the man who has taken the prescribed course in a reputable Osteopathic school is a better and more thorough anatomist and physiologist than any student who has been graduated at a medical school, for the reason that anatomy is made the chief study. The science of Osteopathy being new the student is compelled to undergo a more rigid examination than in the ordinary medical schools.

Osteopathy being based principally upon anatomy the reasonable inference is that the practicing Osteopath understands more about the human frame than the M. D. The senator says the bill is intended as a money-making scheme, and that it originated at Kirksville, Missouri. The senator guesses wrong again. The bill did not originate in Missouri, but was framed by Dr. Harlan, of Denver, who of his own accord and in the interest of Osteopathy in Colorado interested himself in the passage of the bill. That it is a money-making scheme is the most ridiculous assertion that could have been made and shows to what straits the senator was reduced in his efforts to find an argument with which to antagonize the bill. There would not, perhaps, be more than five or ten applicants for certificates during the whole two years of the examiners' incumbency, and these at five dollars each, provided they obtained a certificate, would not prove it to be a very money-making position.

The financial part of the bill was the least thought of when it was drawn by the author. It makes no difference whether the bill provides that a man has studied one day or one hour. The supposition is if he obtains from a reputable school a diploma that he has attended a full course of two years' instruction. The conclusion is inevitable, because no man has ever obtained a diploma without having attended a two years' course and passed a most rigid examination. The school of Osteopathy at Kirksville, Missouri, was not founded for the purpose of grinding out graduates simply for the fee, nor has that institution ever given a diploma to one of its students who was not in every sense worthy of the honor. So strict is this institution that no amount of money could purchase a diploma by anyone who was not entitled to it.

Some institutions, it is said (of medicine), have done this kind of business, but even with them we are generous enough to believe the instances have been very rare. A school of medicine or any other school that grants diplomas for the fee merely cannot do business long among intelligent people. A quack or an incompetent physician will soon make himself manifest in any community where he follows his profession any length of time. "How can an Osteopath," asks the senator, "tell what are obnoxious drugs without having studied medicine?" The Osteopath claims, and his reason is founded upon good common sense, that all drugs taken into the stomach are more or less noxious as well as obnoxious.

The senator finally comes down to the point that if the applicant for a certificate to practice Osteopathy was compelled to go before a medical examining board—a board made up of M. D.'s—then there might be some reason in the bill. Osteopathy has nothing to do with the practice of drugs and the M. D.'s would be about as well qualified to pass judgment on the qualifications of an Osteopath as a lawyer who was familiar with medical jurisprudence would be to pass judgment upon the qualifications of a student of drugs. It is not only necessary that the applicant should produce a diploma from some reputable school in osteopathy, but in addition thereto he is required to pass an examination before the state examiner here, so that while the diploma is prima facie evidence of his possession of the necessary qualifications, he is not entitled to a certificate upon that alone.

Senator Moody, who introduced the bill, followed Mr. Reuter with a most telling and convincing argument showing that Osteopathy was not only a science, but one of the greatest and most beneficial of the present century. He read testimonials to show that thousands of suffering people from all parts of the country and from England had visited Kirksville, Missouri, the home of Dr. A. T. Still, the founder of the science, to be treated by him and that 60 per cent of these had been absolutely cured by him and 95 per cent greatly benefited.

These testimonials might have been largely increased in number by the hundreds of patients who have been successfully treated by Dr. Harlan, of Denver.

The senator said * * * "I do not believe that the strained statements and construction, to put it mildly, of the distinguished and elegant senator from the First should go unchallenged. He bases his arguments upon two propositions. One is that this Osteopathy is not a science. One reason he gives why it is not a science, is, because it is not found anywhere in the dictionary. He claims that it is nothing more than a Swedish movement or massage. It is strange, but true, that a man who professes the greatest solicitude for the people who are to be afflicted by professors of this science should know so little or nothing about it. The senator knows as well as he knows anything that new words are coined yearly. I might cite him to the fact that Emerson coined a number of words in his time, so there is but little in that argument. We have fortunately, some one to help us out upon a definition of science. "Abstract science," Worcester says, "is the knowledge of reasons and their conclusions; science is truth attained by the greatest methodical study, a knowledge of laws, principles and conclusions." Now, Mr. President, before this child had a name, before Osteopathy was born, facts without number that demonstrated the truth of this science were proven by experiment; by course of study, by practice upon the human body, and the name was given, I can tell you how, because I have the document, and as a part of my remarks I shall read from the Journal of Osteopathy, published at Kirksville, Missouri, February 7."

(After reading he continued:

"I trust the senate will pardon me for this. I did not intend to inflict this upon them, notwithstanding there is considerable valuable information in this document. But I believe it is due to senators who are about to cast their votes upon this important measure—important because it will give legal recognition to a science that I am convinced is one of the most important that has been given to us during this century that is so prolific of new sciences.

"By the technical definition of Osteopathy a great many people are led to believe the new science is one which treats only bone diseases or dislocations. The term Osteopathy, like terms that are often applied as names, is not truly indicative of the character of the new method of treatment; in fact it is impossible to formulate a word that will give a correct idea of the system.

"The gentleman from the First made some very remarkable statements with respect to its being similar or practically nothing more than the Swedish movement or massage. He voices, I believe, what he believes to be the truth, and 'as a man thinketh so is he;' he must be truthful. Among those statements was the one that this is not a science, and that the ordinary medical practitioner is a scientific man, and his practice is science. That he should attempt to define what he is pleased to denominate as a science when there is not a medical doctor on the face of the earth, to my knowledge, that has ever dared in this age and generation to declare medicine a science." (Again reads and then continues:)

"When the gentleman couples surgery that is another proposition. All concede that, and none concede it more cheerfully than the Osteopath. I did not intend to say a word on this subject, but the senator's remarks were so untrue that I believed it necessary to say what I have said. The senator would have other senators believe that if Colorado passed this bill it would stand alone in its folly. I would have the distinguished senator to understand that Vermont first led the fight for Osteopathy and recognized it by an enactment. Missouri followed a short time since in this present session to pass a bill by 101 to 16 in the house and 26 to 3 in the senate. Is it fair to assume that

the 101 were as ignorant as the senator would have you believe they must be to vote for this measure?   Is it reasonable to suppose that 26 senators were more ignorant in the
Missouri Legislature than three medical practitioners?
And Missouri is not alone with Vermont.   I understand
five other states have also introduced bills, Michigan and
North Carolina, as well as this state.

"Now, Mr. President, I have no apology whatever to
make for having introduced this bill, and it might not come
amiss to tell you what I did and how I came to introduce
the bill.   There are but two or three graduates in the state
of Colorado; one of them lives in Florence, the other two
in this city.   I want to say to the senators I believe it would
be worth many thousands of dollars to this city to pass
this bill and have an institute where this science can and
will be taught in this city.   Already it has brought to the
city of Kirksville, Missouri, thousands and thousands of
people from all over, I might say, the earth.   We have
testimonials and quotations here from London people.   It
is not necessary to read them.   They come from Canada,
from England, and every other nation of the earth; they
come from every state in the Union."

The senator after speaking of the perfect cure of those
who had been treated at Kirksville, said:

"What other science of treatment of disease or scheme of
treatment of disease can show such a record as that? and
yet the senator undertakes to belittle one of the best
sciences and greatest in my judgment, although so young,
that we have no history of it whatever.   I believe that Dr.
Still's name will go down to posterity blessed by mankind;
they worship his name now and I am glad that that noble
man has the courage to go out and fight the fight that he
has fought and that he is going to reap some recognition
for the valiant struggle he has made."

3

Senator Gallagher: "Mr. President. I just wish to state the experience I have had with Osteopathy. On last Monday night, a week ago, I was taken sick with the "grippe"— and violently so at that. I suffered considerably that night, but the next day I felt a little better. That night at ten o'clock I was taken worse again and during the night and the following day suffered considerable. One of my friends telephoned my wife that he did not think I would live long enough for her to reach the city of Denver. He then said: 'You have got to have a doctor.' I had been taking medicine in order to produce a sweat. I had my friend to call in Dr. Harlan, and in fifteen minutes he removed every pain in my body, and in twenty minutes had produced a sweat over my whole system. That relieved me absolutely from everything, in which condition I continued for two days. In twenty-four hours I was able to get up. I used some other expedient for the same disease five years ago. I got the service of one of those fool doctors, and he treated me for two weeks. I laid in bed and it was a considerable time before I could walk around. I insist at this time that this osteopathic treatment gave me more relief without a drop of medicine, and I have been able to attend to my duties, a little irregularly, but yet in the ring, and I think it nothing more than right and just to recognize a science that has done so much in a case of bone disease, and in a case in which the old school doctor would have kept me in bed three or four weeks.

"I am here to-day and can prove by the other senators who called on me to verify the statement that I could hardly speak. And yet the next day after that treatment I was able to get up and put on my clothes. The third day I came down to my meals as usual, and on the fourth day I was able to get out and take fresh air, and on the fifth I came to the senate chamber and improved every

day so that I feel better now than I did before I was sick.
I do hope this bill will pass."

Senator Crosby: "This is a very important matter that
we are going to pass upon to-day. If it is a good measure
it should be passed, and if we do not pass it we are placing
an impediment on the wheels of progress—either one or
the other. If it is a good thing to pass this bill, and we
do not pass it then we are placing an impediment in the
way of human progress. The senator of the First made
what he thought was a good point when he said there was
only one examiner. This objection can be easily removed
by having two or three or five; that is merely an incident
to the bill. The senator of the First has a method of at-
tacking all these bills that do not conform to past methods;
he is controlled by environment, and is strongly opposed to
innovation. He thinks it must be wrong, and if he were
in an uncivilized country—in the Cannibal islands—there
is no one who would dare to introduce innovations more
than he would. He believes that everything that is is
right, and when you begin to change that must of necessity be wrong. I do not believe in that kind of argument against a bill. He talked considerable about science.
The senator from the Fifteenth talked considerable about
science. They are both wrong. There is a great difference between science and use, and the senator of the First
says he knows nothing about the science of osteopathy.
There can be no such science. Science is defined by Worcester as a knowledge of principles, law, theories, etc., and
use is the application of science. They say Osteopathy is
based, or is endeavored to be based, upon the science of
physiology and anatomy or the science of medicine; the
word medicine is supposed to be based upon it.

"Science can be taught from books, because science is
nothing but theories, the knowledge of relations, laws or

principles, and never can be taught by books, because use is the application of these principles, and you cannot teach the application of principles in books. That is why it is impossible to teach anybody the carpenter's or machinist's trade, or any of these trades by books. Why? Because its application is not expressed in books. Surgery is not a science, neither is osteopathy; osteopathy is use; medicine is use, and it can be applied in either good or in bad ways. The only thing for us to decide then is, is this founded upon science; is this use the application of certain principles of science? If it is then it is worthy of discussion; if it is not then we should brush it aside. Christian science and all these other sciences are based upon no science. They are uses; certain things sought to be done through a certain medium; they are based upon no science; therefore, they are not worthy of discussion; but this use of Osteopathy is based upon science, anatomy and physiology, and I do not care what you seek to cure the human system of, there must be two sciences taken into consideration and those are anatomy and physiology, the science of functions of the human organism. When your organization is not in normal condition there must be either something wrong in the structure or something wrong with the functions, and Osteopathy, as I understand it, seeks by the education of touch to discover immediately where there is anything wrong with the structure; that could be done and I think is done by the Osteopaths, but it seems to me when we endeavor to treat the functions of the human organism in the same way because a function may be deranged, the action of the system may be deranged without affecting the structure of the system in one particular. So it seems to me in that respect, in endeavoring to treat that function of the human organism in the same manner that they intend to reach the structural system.

"Their argument is not based on science. In that particular I am not clear. We are dealing with this new something we do not understand, therefore it must be bad; that is about the argument used. It was but a few years ago that medical practitioners thought when a man had fever nothing could be better for him than to cut one of his veins and let the blood out of him; they thought it was something terrible to permit that man to have a glass of water. Now they tell us the opposite is true. They are continually progressing and adopting something that they have not tried before, and I say when the senator from the First seeks to discredit this use because it is new, because it is something that has not the sanction of the crowned heads of Europe, therefore, it must be bad.

"The same reason that makes it necessary for a state board of examiners, is also necessary to appoint this board, because there are numbers of people to-day who are becoming attached to Osteopathy and its use. These people are going to Missouri to be doctored for something wrong with their systems. If there are doctors of this class in Denver they will come here. The state steps in and says, 'We will protect those who come to our doctors,' and say to them, 'We will guarantee in some particular that these doctors have passed a certain examination and are thoroughly competent to cure that wrong in your system.' But according to the senator from the First the state should say absolutely nothing; the state should not say, 'That man knows his business,' or whether he does or not. This is not a benefit to quacks, but it destroys quacks, and makes them pass a certain examination, showing that they know what they are doing; showing that the state will prohibit them from making experiments upon the human body; showing them that the state legislature has said that when they go to an osteopath the state will guarantee that that man or woman knows his business. That is all there is to it, and it seems to me that the same reason that makes it necessary to appoint a board for the regular practitioners also makes it necessary to appoint this board."

The Colorado Legislature which passed the Osteopathic bill by such a large majority acted in keeping with the ideas of progress, which the people of the state have always exhibited. It was the second state in the Union to adopt woman suffrage, and for many years prior to that time statutory enactments had been made by different legislatures whereby woman stood before the law as man's equal in all the business relations of life. The science therefore of Osteopathy received a cordial endorsement by the people of the state through their representatives, and will, we have no doubt, at some time in the near future, receive the same overwhelming endorsement when another and more generously inclined governor will gladly affix his signature to a bill of that effect. In the meantime the new science will continue to advance and grow in popularity, because it is founded on good common sense and appeals to the reason of every fair-minded and unprejudiced person.

THE GROWTH OF OSTEOPATHY.

The following article was written by Col. A. L. Conger, a prominent manufacturer and a man of national reputation residing at Akron, Ohio. He is now and has been for many years president of the Whitman & Barnes Manufacturing Co., operating plants at Akron, Syracuse, N. Y., St. Catharines, Ont., and West Pullman, Chicago. He is largely identified with the agricultural implement business and served two terms as president of the National Association of Agricultural Implement and Vehicle Manufacturers. He is also a director in the Pittsburg Plate Glass Co., operating plants at St. Louis, Kokomo and Elwood, Ind., Ford City, Tarentum, Creighton and Charleroi, Penn. He is also president of several other corporations. He is prominent in republican politics, has been chairman of the republican state committee of Ohio, and served eight years as a member of the executive committee of the republican national committee. He is vice-president of the Northern Ohio Railway Co. and a director in the Cleveland, Akron & Columbus Railway Co.

"As we read history and learn of the growth and advancement of the different branches of science pertaining to the art of healing we find that each one has grown and advanced in accordance with its worth and merit, as shown by the results secured from actual tests made. Take as a basis that department of the science of medicine which was founded on the theory that drugs are curative. We find that it has been in existence for more than two thousand years and practically without progress, unless the enlargement of its pharmacopœia might be termed an advancement. Be this as it may, there can be no successful contradiction of the fact that for all time as civilization has advanced there has been a constant desire on the part of the higher civilization to break away from drugs in the art of healing. This was most marked when Hahnemann discovered and introduced the system of homeopathy. Hahnemann's trials in introducing homeopathy demonstrated with what terrible force this department of the science of medicine has met any effort to heal the sick without the use of drugs, the very thing which all intelligent mankind were seeking to be relieved. The facts may be further illustrated by following out Hahnemann's theory, which met with such universal favor throughout advanced civilization, and, in spite of the united opposition of the drug doctors, gained and maintained its principles so that every town of any size now has its homeopathic physician. The theory of Hahnemann was to break away from the large doses of medicine, getting down to the attenuation and infinitesimal doses, even going so far as olfaction.

"It is a significant fact that the popularity of homeopathy is greater where civilization is highest, as in the great cities and larger towns. It does not appear that the early popularity of homeopathy was due to the remarkable results obtained by Hahnemann's method of treatment, so much as a willingness of the public to accept the lesser of the two evils.

"In the other sciences, such as orificial surgery, eclecticism, hydropathy, massage and Swedish movement, all of which have merit and intelligent followers, the whole tendency has been to displace drugs with something more reliable in the art of healing. In this proposition the ground was never so fully covered as has been by Dr. Andrew Taylor Still, in the discovery and development of the science of Osteopathy. It is a complete science in itself, and is susceptible, if taken in time, of curing all diseases which have been generally recognized as curable and of greatly reducing the minimum of incurable diseases.

"As I understand Osteopathy from my own observations and personal experiences it is thoroughly scientific, rational and natural. It is founded on a logic, a philosophy and a constructive basis that will bear the closest investigation. The Osteopathists from the nature of their work must be, and I believe they are, the best anatomists in the world. This will appear only reasonable when it is remembered that the whole practice is founded on a most thorough knowledge of all the parts and processes of the human body in health and disease and that there is not a single osteopathic operation, treatment or manipulation that does not require the most exact anatomical, physiological and pathological knowledge to perform. As regards anatomy and physiology this is true of no other method of healing except, possibly operative surgery. While anatomy and physiology are taught in medical schools, unless the medical graduate is going to be a surgeon he is apt to make no practical use of this knowledge, and it soon slips away from him. The Osteopath who would allow his knowledge of anatomy and physiology to slip away from him would be in as bad a predicament as the medicine doctor who would go to see his patient without his pill-bags.

"As a therapeutic agency, Osteopathy is working revolutions in the art of healing. In the work it has performed at the Still Infirmary, at Kirksville, in the earliest stages of its introduction to the world, it has made for itself a record of results along the whole category of diseases, which certainly challenges the respect and admiration of all other departments of medical science. The building now occupied by the A. T. Still Infirmary was opened for patients in January, 1895. Since that time two new wings have been added increasing its capacity about three-fold, and about seven thousand patients have been treated. The great mass of these patients had been pronounced incurable by many of the arts of healing, and yet more than fifty per cent of such cases have been absolutely cured, while a minority of the remainder were greatly benefited, and a case is seldom found in which no benefit is received. All this has been accomplished without advertisement. Patients who have gone to the Still Infirmary for treatment were universally induced to do so from the favorable reports of friends or relatives who had visited the institution.

"While Dr. Still has been at work on this science for more than twenty years, it has only been brought before the public in the last ten years, during which time public attention has been called to the many remarkable cures made by Dr. Still and his assistants. Osteopathy has made rapid strides and established for itself at this infirmary and at other points, cures, a record of which, certainly merits the unbiased investigation of the American people and thinkers throughout the world.

"The infirmary to-day has a capacity of treating five hundred people daily, while the American School of Osteopathy is equipped to accommodate one thousand students. The growth of this school has been simply wonderful. One year ago there were only fifty students in the school, while to-day there are over three hundred enrolled.

"This shows the enormous increase of 500 per cent in one year, and is a fair index of the growth of osteopathy during the year. One third of these students are ladies. The science opens a new field for women and they are destined to become experts in the art of healing by Osteopathy. In fact, looking at the matter from a business man's standpoint, I believe the science of Osteopathy affords an opportunity for intelligent, ambitious young men and women that is not equaled at the present day in any art, trade or profession. The study is fascinating, the work is pleasant and the results are of such a character that the osteopathist will always be paid handsomely for his services.

"The real growth and popularity of Osteopathy among the people can best be shown by calling attention to some of the contests for recognition before our state legislatures. These bodies are chosen directly from, and their members are in closest touch with the people, many of the great and important questions being discussed at the family firesides. The great medical trust which has controlled and manipulated legislation in all the states for many years past, touching the science of medicine, have seen to it that all avenues were guarded and none but allopaths or drug doctors are protected by state legislation. They are installed in all institutions and places regulated by law. Therefore, when the graduates of the American School of Osteopathy came knocking at the doors of the state for the right to practice the newly discovered science they found themselves shut out from all legal recognition, without permission to practice in any state. Dr. Still himself being a regular M. D. could not be shut out under the law. He could found his science, but was handicapped in developing it. So that two years ago a bill was introduced into the Missouri Legislature recognizing the American School of Osteopathy and regulating the practice of this science in the state of Missouri.

"After much discussion the bill passed both branches of the legislature, but the influence of the medical trust was too powerful for Governor Stone, who vetoed the bill on the plea that osteopathy was a secret science. The vetoing

of this bill amounted practically to the referring of the
whole question back to the people, who for a second time
were to become the real jurors and arbitrators in this
case. Osteopathy in the meantime continued to grow in
favor, and became a public science, not only with the people
of Missouri, but in the whole northwest, and wherever it
was given an equal chance to compete with its enemy,
drugs. Last year a new governor and legislature were
chosen by the people of Missouri. In view of Governor
Stone's veto, Osteopathy in many localities became the
paramount public issue.

"It had grown in favor with the people, so that in this
contest it gained a complete victory. Governor Stone was
not only beaten for renomination, but the new legislature
was overwhelmingly elected in favor of recognizing Osteopathy. A new bill was introduced into the Missouri Legislature legalizing and regulating this practice, and the
'secret science' had become so public and popular that the
bill passed almost unanimously. Thus the will of the people and the growth of Osteopathy were triumphant in Missouri, the home of the discoverer and founder and the
parent school. All this was won upon just and true merit.
While this struggle for Osteopathy in the state of Missouri
was going on, a bright young graduate of this school, Mr.
George J. Helmer, had commenced to practice and demonstrate the wonderful cures of the newly discovered science
in the state of Vermont. So great was his success that the
medical trust of that state became alarmed and were the
aggressors in forcing the issue against Osteopathy. They
sought to forestall the work of this science by the passage
of a law preventing its practice in the state of Vermont.
The friends of Osteopathy, however, rallied at the state
capital, and when the final hearing was had before the several committees and members of the Vermont Legislature

the bill presented by the drug monopoly was overthrown
and an entirely new measure legalizing the practice of Osteopathy in Vermont was introduced in its stead. In one
hour and fifteen minutes this bill was passed, signed by
the governor and became a law. Thus Osteopathy, purely
upon a showing of its merit and popularity with the people
won in the state of Vermont. Next came the now famous
fight of the able, brave and courageous Mrs. Helen de
Lendricie, of Fargo, N. D., for the recognition of Osteopathy in that state. Through personal experience recounted
elsewhere in this issue, she had become convinced that
Osteopathy was a science, which the people of her state
ought to have, and upon her return home from a visit to
Kirksville she began the campaign single-handed and
alone. Delegations of doctors from the larger towns in
North Dakota were at the capitol when the legislature con-
vened to fight the admission of the new science, but, as
she expresses it, 'one woman and mighty truth won the
day.'

"The next legislature to recognize Osteopathy was that
of Michigan. Many prominent people from that state had
visited Kirksville and made careful investigation of the
newly discovered science during the past year. The result was the introduction of a bill legalizing and regulating
the practice of Osteopathy in the state of Michigan. The
bill passed the senate by a vote of 24 for to 1 against, and
the house 72 for and none against. It was signed at once
by Governor Pingree, who was much in favor of the measure, having given it a careful investigation. As Osteopathy
gains recognition in the different states and thus breaks
down the gigantic medical monopoly it renders incalculable
aid to the poor man by securing to him the benefits of
competition in the healing art, and as Governor Pingree is
known all over the country as the poor man's friend, op-

posed to trusts and monopolies, he would naturally be for
any just measure that would aid this class. In Colorado
and South Dakota similar bills were passed by the legislatures, but were vetoed by the governors, who, like Governor Stone, of Missouri, believed in referring Osteopathy
back to the people. In North Carolina the Osteopathic
bill passed both branches of the legislature, but did not become a law, owing to some technicality. In the state of
Illinois, where an Osteopathic bill is now pending, it has
passed both houses by an overwhelming majority, but has
not yet been signed by the governor.

"In all these contests for recognition before the law in
different states Osteopathy has asked for itself no exclusive privileges. The only protection it needs is that of
compelling its students to spend the proper length of time
in its regular schools so they may become able and competent practitioners of the new art and to shut out impostors
from among its own ranks. This science seeks only a fair
trial and an opportunity to win its way to the front by
actual merit and the favorable results it may obtain.

"From the foregoing it will be seen that while Osteopathy
has made rapid growth before the people of the country
it has met with corresponding recognition before the legislative bodies, having been legalized in four states during
the past year.

"It is charged that those who write or speak of Osteopathy are perhaps over enthusiastic; but it is true that the
science of Osteopathy performs most wonderful cures, and
these cures coming under the personal observation of those
who speak and write on the subject cannot help but make
them justly enthusiastic. Take the case of the writer.
He was stricken with paralysis while in Boston, on the
11th of January last, arrived at this infirmary in Kirksville on the 17th of January in a helpless condition, his

whole left side being paralyzed. He was carried to his boarding house and placed under Osteopathic treatment. In four weeks his recovery was such that he could arise, dress himself and walk about. If such recoveries and such results of Osteopathy are not calculated to make patients enthusiastic it is hard to tell what would. But while some of these cures seem like miracles they are simply scientific, and science becomes less marvelous when better understood."

THE OSTEOPATHY CHASE.

"Dr. Ruggs and Dr. Muggs went forth one summer day,
　To listen to the robins' song and talk along the way.
Their instruments and pill-bags too, in lusty hands they
　　bore,
　And for fear of snakes and sich, they rolled a barrel
　　before.

"Said Dr. Ruggs to Dr. Muggs, who was very short and fat,
　'I wish you'd tell me what the deuce a man is driving at
When he affects to scorn our drugs and instruments you
　　see—
　And claim that Still can cure each ill with osteopathy?'

"Said Dr. Muggs to Dr. Ruggs, who was very tall and lean,
　'It is not right for Still to fight us, confound him, it is
　　mean,
With pills for chills and brace to make a limb that's weak-
　　ened stout,
　And powders, tonics, forceps and knives, we know what
　　we're about.'

" 'True we don't so many cure, but then folks think we
　　do—
　And it's all the same to them you see, likewise to me
　　and you.'
Just then loud thunder struck their ears, and horses
　　came in view,
　With upright riders on their backs, and lassos strong
　　and new.

" 'The Osteopaths, the Osteopaths,' cried Muggs in awful
　　fright,
　And fanned his hot and florid face, while he com-
　　menced his flight.

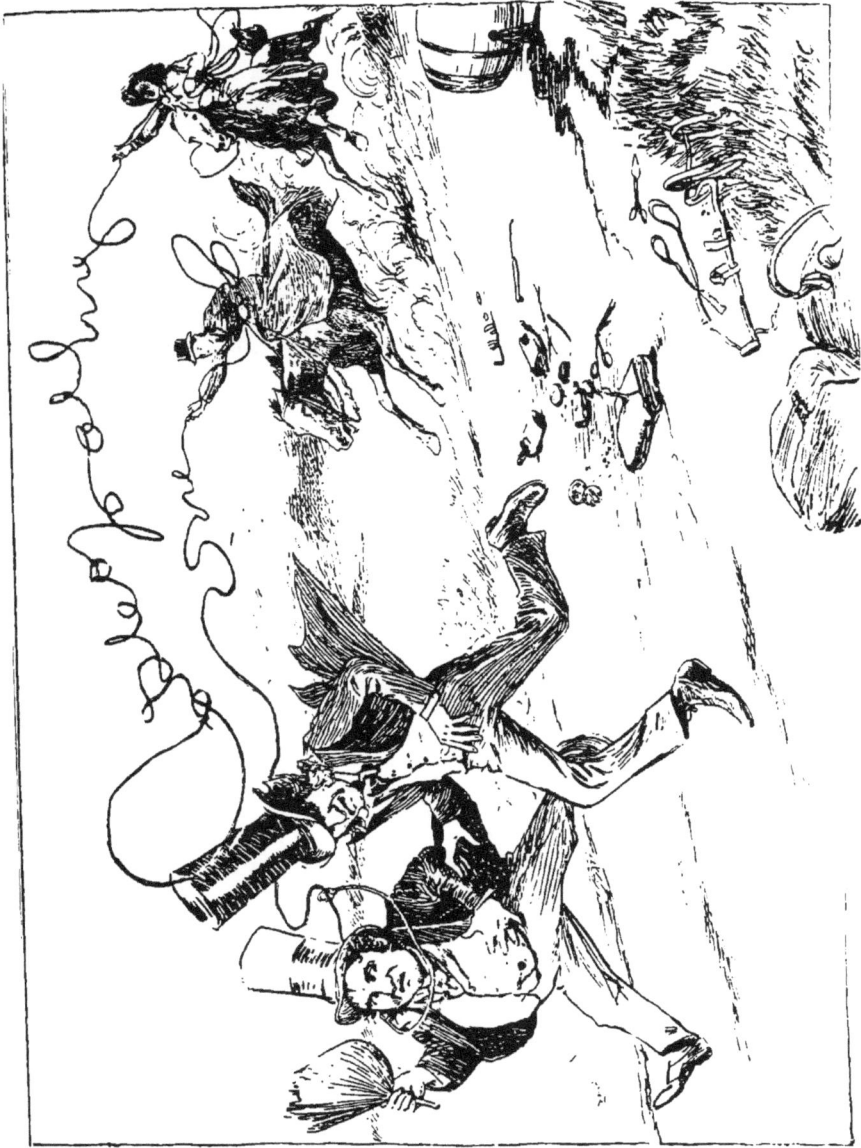

THE LASS CATCHES THEM.

Ruggs ran too, close at his side, his soul filled with alarm,
    The Osteopaths to still their fears cried out, 'We'll do
        no harm.'

"On they ran, dropped brace and truss, and pills from pill-
        bags flew;
    Scissors and knives, all medicines, and likewise forceps
        too.
The Osteopaths were gaining fast, and loud with mirth
        they roared,
    While from the bung of the whisky keg, the bracing
        tonic poured.

"The lady horseman roped in Muggs; her companion
        threw you see,
    And the lasso whirling through the air, formed Osteop-
        athy.
Thus Osteopathy has caught those doctors Ruggs and
        Muggs,
    And always wins when for its foes it has old death
        and drugs."

OSTEOPATHY AND LEGISLATION.

From the Journal of Osteopathy.

"When Abraham Lincoln issued his emancipation proclamation, he simply carried into effect in one direction the spirit of the Declaration of American Independence and the Constitution of the United States. The wording of these documents is so very plain as to leave no room for doubt as to the intention of their framers. They meant to found an absolutely free country, the laws of which should in fact extend equal rights to all and special privileges to none. Yet the people of this country have been from the beginning and are now under a bondage that makes them anything but a free people. They are under the absolute domination of drugs in the hands of the so-called 'regular' or allopathic school of medicine. This school was handed down to us from the mother country, and very early in our history secured a franchise for the healing of the sick, and has so fortified itself by legislation as to make that privilege practically an exclusive one. As soon as a state was organized and admitted into the Union the representatives of this school proceeded at once to procure the passage of such laws as would protect them and exclude all others. It was but natural that the framing of medical laws should be left to the medical members of the legislative bodies, and they were not slow to take advantage of the opportunity thus afforded them. Thus there was established a monopoly that was more fully protected by legislation than any of our modern trusts.

"The allopathic school of medicine was founded upon the supposed curative properties of drugs, and while it was found necessary to modify its original claims and plans to some extent, dosing with drugs is still its fundamental idea and is made the basis of all medical laws and rules and regulations of the boards of health existing under those laws. In almost every state in the Union we find all power and authority in all matters pertaining to healing vested in the representatives of this system, and where other schools of medicine are admitted at all, it is never on an equal footing with the old 'regulars.' Even the doors of the army and navy are closed against every other system and all advances made through any other channel. Having such full control they can and do assume the right to make all advancements and discoveries in the science of healing; and every discovery made, whether inside or outside of their ranks, must wait until they approve and promulgate it. An iron clad code of ethics binds them together, any violation of the provisions of which is unprofessional, and unprofessional conduct is, under the law, sufficient cause for revocation of license to practice.

"But the order to wait could not always be obeyed, because of the fact that the people. are always looking for something better, and care little for 'regularity' in such matters. Thus, Homeopathy, Eclecticism, Swedish Movement, Massage, Psychotherapy, Electropathy, Hydropathy, Gastropathy and many other pathies have compelled the old school to extend to them more or less recognition. The 'regulars' have appropriated some of the systems and improvements outright, after having ridiculed and reviled their believers. In some cases this adoption has been only after long delay and then under different names from those originally given the new system. The Homoeopathists and Eclectics have secured some independent legal standing but have been compelled to fight hard for it, and have to be content with the simple permit to practice, the old regulars reserving unto themselves practically all of the state

and federal patronage and support. Missouri has just
witnessed an exemplification of this in the difficulties encountered by the governor in attempting to place one of
the state insane asylums under homeopathic control.

"Now comes Osteopathy, a science of healing without
dosing with drugs, complete within itself, not needing the
support of any of the drug schools. It has stood alone
for over a quarter of a century, and has demonstrated its
ability to materially reduce the death rate in all curable
diseases, and successfully handle many diseases heretofore
pronounced incurable; but, not being on a drug basis, it
could secure no legal standing unless it formed an alliance
with some of the systems founded on the drug theory. Dr.
A. T. Still, the founder of Osteopathy, recognized that any
merely expedient combination would prevent the proper
development of his discovery, so he kept it free from any
alliance with drugs and developed it on an absolutely independent basis. The position occupied by the science at
the present time, as a result of his policy, amply proves the
wisdom of this course. Being himself a registered practitioner of the regular school he could not be prevented
from practicing as he chose, but when he organized a
school and began teaching his science to others who were
not medical graduates, and demonstrated that his system
was founded on scientific principles and could be taught,
he encountered the combined opposition of all the representatives of the drug schools, who sought to close up his
work and drive him to introduce drugs into his system.
He was often discouraged, and at one time closed up his
school because of the legal obstacles in the ·way of his
graduates practicing, but he never wavered from his purpose of keeping Osteopathy pure and free from debasing
alliances. Osteopathy was, however, too great a boon to
mankind to be thus lost, and he was induced to open the

school again by the strong pressure brought to bear by the people who wanted the system perpetuated, and people who wanted to study it, and who were willing to take their chances with the laws. Under such circumstances, the progress of the school was necessarily somewhat slow at first. Those pupils who were brave enough to enter the work in the face of these obstacles had that vigorously independent spirit needed to aid in pushing the work forward. Many prosecutions were instituted, a few convictions resulted, but the penalty for violation of the law was of such a nature as to permit the accused to demand a jury trial, and a verdict of acquittal was usually rendered. In the very nature of things, prosecutions were seldom begun until after the Osteopathist had secured some results, curing a few cases that were not yielding to former treatment, thus interfering with the local prescriber of drugs, but these results arrayed the people on the side of the accused, and obnoxious laws were hard to enforce under such circumstances, there always being a doubt as to their applying to a system not using drugs or the knife.

"Most medical laws are so worded as to impose their conditions upon all practitioners of medicine and surgery, but usually go on and attempt to include thereunder all who pretend or attempt to cure disease by any means whatsoever, although, of late years, some few of the more progressive states have so modified their statutes as to make such laws apply only to those who use drugs or the knife. In the broad sense the terms 'medicine and surgery' would embrace all remedial agents and measures, and the attempt has been made to so construe it, as applied to Osteopathy, but it has never yet been done. Most of these iron clad laws are so arranged that could their literal rigid construction be enforced, all nurses, midwives, bath establishments, massage and Swedish movement cures, and all who

attempt to alleviate human suffering by any means, would
be excluded. However, as long as a system does not materially interfere with the practice of the old regulars, no
trouble is made, but when the Osteopathist comes into the
field and cures some cases that the disciples of drugs have
pronounced incurable, then the weakness of human nature
usually asserts itself, and the attempt is made to drive him
out. It cannot be claimed that no medical laws are needed.
There is good reason for suitable restrictions being thrown
around the practice of the healing arts, especially where
drugs and the knife are used, and standards of qualification
should be fixed, but the argument that the people need
protection falls powerless in the face of the fact that no
very great proportion of them admit the necessity of having any law providing that they shall or shall not employ
any person or system to treat them or their families when
in sickness. They usually feel justified in denouncing
such measures as unwarranted interference with their personal liberties. Besides, almost every page of printed matter, every available stone, board, fence, wall and roof within reach of the eye of the passer by, sets forth in glowing
promises the virtues of the many nostrums and remedies
that are on sale at every street corner and cross roads store
in the land. Every drug and poison used by any physician
is thus placed within the easy reach of everybody at much
less cost than it can be prescribed by the physician. However, the Osteopathists could not feel called upon to revise
and remodel the entire medical laws; they could not hope
to succeed if they did undertake to do so.

"In view of all these circumstances it seemed wise that Osteopathy should ask for direct, independent recognition at
the hands of the immediate representatives of the people,
the law-making powers. The states of Missouri, Vermont,
North Dakota and Michigan have passed special laws recog-

nizing the science and granting its graduates the right to practice, under suitable restrictions. In Ohio, the judiciary has sustained the rights of the Osteopathists to practice in that state. In some other states the laws have been so modified as to permit the practice without special recognition. Some states have repealed all laws restricting the practice of medicine, leaving the field open to all.

"In almost all cases where changes in the laws have been made in order to permit the practice of Osteopathy, the demand has come from the people of the state, because they want Osteopathy, and in order to get it, go to work independently to secure the removal of the barriers without any help from the representatives of the science. Every genuine has its counterfeits. There are many, and will be more, counterfeits, pretenders and frauds who will use the good name of this science to humbug the afflicted public, it being so new and known to so few, comparatively, that the people have no means of knowing the genuine from the counterfeit. Therefore, the legislative bodies that have acted in the matter up to the present time have seen fit to place some restrictions around the graduates. No one can raise serious objections to this as long, at any rate, as the present policy of regulating such matters by law is kept up.

"Osteopathy asks no special privileges; it desires only the right to be used if the people want it. It seems but fair that it should have that right in every state in the Union, and as it becomes better known the people will no doubt demand that they shall have it, and our national congress may be asked to give the science a trial in the army and navy. Its friends believe that if introduced into the army and navy it would greatly reduce the death rate."

"A BRILLIANT LEGISLATIVE VICTORY.

From the Journal of Osteopathy.

"One of the most distinguished friends of Osteopathy is Mrs. Helen de Lendrecie, of North Dakota. She is the wife of the 'Merchant Prince of the Dakotas,' Mr. O. J. de Lendrecie, of Fargo, who is one of the best known business men in the northwest.

"Mrs. de Lendrecie was in Kirksville last fall for treatment, and was cured. While here she made a thorough investigation of Osteopathy and the work it was doing. Immediately upon her return home, she had a bill framed legalizing the practice in North Dakota, and alone began a campaign to secure its passage.

"The measure met with decided opposition. Delegations of medical doctors from all over the state congregated at the capitol early in the legislative session and did all in their power to kill the bill.

"Osteopathy was almost unknown in North Dakota, and it required a campaign of education to arouse any interest in the subject, but Mrs. de Lendrecie was equal to the task. After many weeks of heroic work, the bill pased the senate. The opposition then concentrated its forces in the house, and especially before the house committee, to whom the bill was referred. The committee was composed largely of doctors, and when the final vote came only one member favored the bill. This looked pretty blue for Osteopathy, and in the ordinary course of legislation would have been the end, but Mrs. de Lendrecie was not so easily discouraged. Her firm conviction that the cause was right, and her most extraordinary determination to triumph, are best told in her own words. In relating an account of her fight,

she recently said of this occasion: 'I was greatly depressed
at the outlook, and it seemed as if my fight for Osteopathy
must end in dismal failure. I called to mind the trials of
Dr. Still in the early years when he was striving to develop
this science. I pictured him as climbing up a mountain
path, with his wonderful eyes fixed on Truth at the sum-
mit, with no light to guide but the reflection of her face
upon his own as he steadily mounted upwards with her flag
in his hand. And then I thought, I too am only a flag-
bearer in one of Truth's great squadrons, but I will wave
it in victory over my head, or go down in the battle to
defeat with it still in my hand; I will never surrender my
flag. The next morning when friends told me that my
case was almost hopeless and advised me to wait until
another session and obtain more help, I asked: 'Have I
any chance?' They replied: 'One man stands pat; you
have one chance in a hundred.' 'Then friends', I said,
'give me one chance in a thousand and I will win.'

"Mrs. de Lendrecie succeeded in having her bill called up
out of general order, and then the speaker, who was friendly to the cause, gave her the desired opportunity of ad-
dressing the members. When it was known that she would
speak, the galleries were crowded, and the senate adjourned
and came into the house in a body to hear her. The occasion was intensely dramatic. The delegations of doctors
were there, carefully guarding their fences and determined
that this new pathy should not break into the green pas-
tures of North Dakota. Mrs. de Lendrecie was escorted to
the speaker's stand, and, as said by the Dakota dailies,
made the 'most persuasive and effective address ever heard
in the state.' She reviewed the arguments that had been
made against her bill and showed their absurdities, and
with incisive irony and irrefutable logic, couched in re-
spectful but plain English, she stripped the mask from the

great medical trust and let the law-makers of North Dakota see this arrogant monopoly as it is, divested of its
'professional dignity.' As she warmed up to the subject
it was apparent that the supposed defeat was being turned
into a brilliant victory, and when she sat down amid
deafening applause, all the plug-hatted and gold-spectacled
medicos in Christendom could not have prevented the
passage of the Osteopathic bill. The doctors, who had felt
so secure only a short hour before, now sat mute and be-
wildered, as though glued to their seats. A vote was
ordered, and the measure passed by a good round majority.
The Governor signed the bill within a few hours of its passage, and it will become a law July 1st. Thus ended a
legislative achievement that is absolutely without prece-
dent.

"Mrs. de Lendrecie is a woman of fine personality, magnetic manner, commanding carriage and finished culture.
She has an active, analytical mind, quick to grasp and hold
a truth; she is a forceful writer, and in social life is popular
and much esteemed. Her espousal of the cause of Osteopathy was inspired by the noblest of motives—gratitude for
benefits received at its hands and a firm conviction that in
fighting for this new system she championed the cause of
truth.

"As has been stated, she became interested in Osteopathy
through her own cure at Kirksville. The facts in her case
may be of interest to the reader, and in order that there
might be no misrepresentation, the editor requested Mrs.
de Lendrecie to write an account of her experience with
Osteopathy, which she has kindly consented to do. Following are the particulars in her own words:

" 'Editor Journal of Osteopathy:
" 'You asked for the reasons which induced me to seek
Osteopathic relief. I will tell you as briefly as possible.

In the fall of 1895 a lump appeared in my right breast. Our family physician advised its immediate removal, assuring me that nothing but the knife could remedy the evil, and stating that it would soon assume a malignant form if not removed without delay. Knowing him to be a fine surgeon, as well as physician, I placed myself in his hands and submitted to an operation whereby my entire breast was removed. It was a great shock to my nervous system, and I had not recovered from it, when the same trouble appeared in my left breast. I had heard meantime of Osteopathy and resolved to try it before again submitting to the knife. However, in September, 1896, I was examined by a specialist in Chicago, who declared that only the knife would remove the trouble, as in the previous case. Instead of submitting to another operation, I went to Kirksville and was completely cured in six weeks' time. My own eyes saw and my own hands felt the obstructions that caused the trouble in both cases, and I know very well that the knife was never necessary in my case. I do not want to be understood as denying the use of the knife, however, for in some cases, I am sure it is necessary to prolong life. What I object to is the haste with which it is called into requisition. I never believed in drug medication, but surgery appealed to my reason. I have perfect faith in the integrity of the man, and the skill of the surgeon who operated upon me. I believe he did by me as he would have done by his own wife, and if I was in need of surgical aid to-day, he of all others would be my choice to do the work. That, however, does not prevent me from declaring that I was the victim of unnecessary surgery, and I am sure the knife is used ten times when unnecessary to one time when necessary. A surgical operation is a dreadful shock from which I believe the system never fully recovers. Osteopathy has clearly proven its right to recognition in the healing of cases heretofore declared only curable by the knife, and it is only right that its supporters should sustain its claims. I know what it has done for me, and I am now and ever expect to be the firm friend and loyal defender of Osteopathy.

"'HELEN DE LENDRECIE.

"'Fargo, N. D., May 23, 1897.'"

SCOPE AND SPIRIT OF OSTEOPATHY.

As a preface to what may be said in the following pages, on the science of Osteopathy, it would be impossible for us to present a more concise, terse, and intelligent statement of its scope and spirit than that which has been given by Dr. A. T. Still, and which has been laid down under eight different heads. These comprise in a nutshell the basic principles upon which the new science rests, and show its present importance, and the possibilities to which it may attain.

"First: Osteopathy, from its inception to its present position of development and success, has been and still is a science—that is, it is based upon exact, definite and verifiable knowledge of the anatomy and physiology of the human mechanism, including the chemistry, histology, morphology and psycho-physics of its known elements, and such reasoned conclusions from this practical knowledge, as make discoverable the great laws of the human system, by which, nature apart from artificial and medical stimulation, may recover from displacements, disorganizations and consequent disease, and regain strength and health.

"Second: The science of Osteopathy consists in the profound study of these facts, forms and forces of the human organism, under all attainable conditions of literary research, morbid anatomy and normal life; and so regulating and controlling the laws of each organ, according to its original and normal structure and function, as to secure and maintain the natural equilibrium of health.

"Third: Osteopathy has, therefore, an immovable basis in nature itself, and that its operations are in harmonious accord with the ineradicable and irrepealable laws of nature, and that its future both in scientific achievement and remedial results, is as illimitable as the boundless and inexplorable resources of universal life.

"Fourth: Osteopathy views man as a microcosm—a miniature of the cosmic universe—and that the proper study of man as a passive machine and as a living organism, opens up radiating lines of research into all the departments of thought and of things, and relates itself naturally and logically to all the great sciences both of the organic and inorganic world; and it furnishes a new organizing principle, by which many of the facts of these sciences may have an entirely new interpretation. Its study, therefore, is most ennobling to intellect and feeling; it is enriching in wisdom to understand, and empowering in ability to mitigate the ills to which flesh is heir.

"'Know thyself.
Enough for man to know,
The proper study of mankind
Is man.'

"Fifth: Osteopathy is in no way affiliated with pharmacy and medicine, except as the effects of these may be known to be avoided. Osteopathy runs a line of cleavage through the entire so-called 'History of medicine,' and divides it into the facts of anatomy, physiology and hygiene on the one hand, and the facts of pharmacy and chemistry on the other. The original practitioners were anatomical and physiological; the chemical or medical practitioners were irregular. So Osteopathy can show from history, reason and nature, that the doctor of medicine is still irregular, and that the doctor of Osteopathy alone is regular. The scholarship of the medical fraternity is challenged to deny this distinction. Osteopathy is a science; medicine is not, and never has been, and all its doctors can not show that it is.

"Sixth: Osteopathy appeals from first to last to facts. Nothing tells like results. Facts are the biggest forces that rule the world. Fiction cannot be foisted upon a practical public as fact. The minds of to-day are impatient with mere theories and speculations. They clamor for facts. They ask for results and returns. Osteopathy does not evade the challenge of a critical and curious public. It courts investigation. It is not afraid of all reasonable tests. It stands only on its merits. It does not ask any one to bankrupt his reason, and then pension him on a miserable allowance of faith. Its principles and philosophy are as capable of illustration in disease as are the rules of mathematics in numbers, or of forces in mechanics.

"Seventh: There is no culture of character, refinement of feeling, brilliance of intellect, keenness of reasoning, no polish of manners, completeness of education, grace of literature, resources of scholarship, or ambition for discovery, that may not find free and full exercise and expression in the proper study and practice of Osteopathy.

"Eighth: It has the peculiar power of infusing the deepest interest into its students, and enthusing the most phlegmatic of its patients with the spirit of love and loyalty for its methods and results. It possesses the wonderful charms and fascinations of nature itself. In its fine and finished form, with scholarship, literary grace, logical power and scientific spirit as its aids, Osteopathy will fast win its way into all the world. Already it has hosts of friends among the great and good of our land, from the new president of the United States to the lowliest citizen; and it is not unknown across the sea."

DR. A. T. STILL.

THE STRUGGLE FOR RECOGNITION.

Had Dr. William Harvey been a man of putty, or in any
sense, a week-kneed supporter of what he conceived to be
the fundamental problem in physiology, he never would
have discovered and proved his theory of the circulation
of the blood. His theory was not only bitterly opposed,
as being a mere chimera of the brain, but so little was he
appreciated, that his prescriptions among the apothecaries
of his day were regarded as entirely worthless. He continued however, for years, in spite of all opposition, to prose-
cute with a sleepless purpose, his theory of the circulation
of the blood. He laid down the proposition that wise men
must learn anatomy, not from the decrees of philosophers,
but from the fabric of nature herself, and in pursuance of
that idea, his investigations covered a wide range of subjects, extending from man down through the animal
kingdom, including the unhatched chicken, and even the
serpent, before he announced the complete demonstration
of the problem for which he had for so many years labored.

When Edward Jenner promulgated the fact that he had
discovered vaccination, and could explain the relation
that cow-pox bore to small-pox, he was opposed by the most
noted physicians of his time. He went to the city of
London, for the purpose of demonstrating the truth of his
theory, but after remaining there three months, and finding no person who was willing to be vaccinated, he returned again to the country, the doctors in the meantime,
declaring that Jenner's theory was not only preposterous,
but dangerous to men as well as to animals. Jenner however, nothing daunted, realizing the truth of what he had

so long advocated, succeeded in finding a patient who was willing to run the risk of a vaccination. The test was made, and Dr. Jenner's name became immortalized, not only in England, but throughout Europe.

Galileo under the threats of enemies, and the menace of torture, still secretly held to the Copernican theory of the solar system, and finally demonstrated to his skeptical contemporaries that his theory was true; a theory which is the accepted one of even to-day.

Had Newton listened to the philosophers of his time, he never would have revolutionized the old theories of the laws of gravitation. He dared to say the world was round, and tenaciously clung to this idea in the face of ignorance and prejudice. His enemies were many and powerful, but all their boasted wisdom, all their personal influence and opposition, as great as it was, did not deflect him a single hair's breadth from the conclusions he had formed concerning the laws of gravitation. He knew that he was right, and all others were wrong, and to-day his name shines as one of the most brilliant stars in the firmament of the greatest scientists.

Had there not been a Columbus, the discovery of the New World might have remained merely a problem, for an indefinite period. Its solution would have certainly not been attempted by any of his contemporaries, for they, the best, and most accurate geographers of that age, scoffed at the theory he advanced, and prevented him by their well-known opposition, from obtaining an opportunity to prove what he claimed for many years after the thought had been projected. It is a well authenticated historical fact that the story of Columbus, his trials and difficulties, have been but the trials and difficulties of all the men who possessed an advanced thought, and proclaimed it to the world. There have been, among a few of the scientists, those who thought they knew all that was worth knowing along given lines, a kind of "dog in the manger" policy. They were failures themselves, and so jealous were they of the reputation and success of others that they did all they could to handicap and throw stumbling blocks in their way.

Scientific men pronounced Robert Fulton's scheme visionary and impracticable, but he went on with his work. Years passed before the realization came. It was on the 10th of May, 1807, that a large concourse of people assembled at the wharf on the banks of the Hudson river, to see the steamer "Clermont" leave the shore, propelled by steam and destined for Albany. The more ignorant of these were guided thither by curiosity. The great majority met there as scoffers to give vent to their shouts of derision, believing that the steamer would prove to be a signal failure, and that Fulton's ideas, which he had cherished for years, would be exploded in the failure of the invention upon which he had staked his fortune and his name. But behold the results which have followed from that successful trial trip of the "Clermont!" Look at the river and the ocean steamers, which have since plowed all the navigable streams of the habitable globe, and have carried their freightage of humanity and commerce into every part of the world, driven by this same agency, that was thought out, developed and applied by Robert Fulton.

Charles Goodyear, to whose inventive genius the world is indebted for the thousand and one useful articles which have been fashioned from India rubber was regarded as a dreamer and a monomaniac. He was twice arrested for debt, and in the desperate straits to which he was reduced, pawned even some of his wife's trinkets. He passed through the fiery furnace, but so absorbed was he with the thought, and so convinced was he, that India rub-

5

ber could be widely utilized for the benefit of the human race, that he never wavered or flinched until he demonstrated that fact to the world. The overshoes, and rubber overcoats that we wear, the life-savers that are used on ships at sea, the hose with which we sprinkle our lawns, the tires that play such a conspicuous part in the bicycle business, and the unnumbered uses to which India rubber is devoted, are all the result of Goodyear's inventive genius, for he suggested the first thought for its use.

Eli Whitney, a poor, unknown school teacher, of northern birth, emigrated to Georgia, and conceived the idea of the cotton gin, an invention which has done more to increase the acreage of the cotton fields, and to expedite the preparation of cotton for the mills, than anything which had theretofore existed.

Elias Howe, while laboring for his daily bread, for the support of himself and family, dreamed and brooded over his invention until the thought was practically embodied in the sewing machine which bears his name, and which is to-day the greatest blessing of which womankind has been the recipient.

Cyrus McCormick, whose reaper has proven such a boon to the agriculturists of all lands was time and again discouraged by friends, who tried to persuade him that the efforts were only a waste of time. To-day the McCormick reaper does its work not only upon all the continents where farming communities live, but it is seen upon the islands of the sea.

Had Professor Morse taken the advice of friends, the progress and development of telegraphy might have been stayed for years, and civilization checked in its onward progress. The old world and the new would have remained upon the opposite sides of the ocean, and had no communion with each other, save through the old medium of ships. The possibilities of the Atlantic cable would have been a dream, and the world would have been deprived of the innumerable blessings which have been conferred upon it, by the use of the telegraph. Intellectual growth would have been retarded. Commerce would have been slow in development; business and trade would have lagged, and all the interests of men, as well as of governments and nations, would have moved along in the same old grooves.

The man who attempts to invade the field of science, with the purpose of adding another trophy to those which have already been won, will not find his pathway strewn with flowers. But upon the contrary, the road is rough and rocky, and at times will be found to be almost impassable. A portion of the way may have been blazed by those who have preceded him, but, if he is to win in his researches, he must pass the uttermost limits which have hitherto been attained by any and all others. He must extend the boundary lines, and cover entirely new territory, which has never before been trod by any of his predecessors or contemporaries. He must become the solitary adventurer, upon virgin soil; upon soil which has never been pressed by the foot of man. To do this, he must be equipped with a little better and more thorough knowledge than any of his contemporaries, concerning the means, which contribute to the end he proposes to reach. In addition to this, he must possess an abiding faith in himself, and his own resources, and be so determined of purpose that no discouragements, however appalling, will deter him from the achievement of that upon which he has set his heart and bent his energies.

The man who dares to combat the old established theories or dogmas, which have been advocated and upheld by learned universities, with a long line of learned pro-

fessors, who have been willing to travel right along in
the old, worn ruts, not only runs the risk, but is generally
branded as a fanatic. Indeed, if all men who have revolutionized the world of thought, by some startling theory,
could so easily get off with no more opprobrious epithet
than "fanatic," they would be especially favored. Not a
few of them have been regarded as being, mentally speaking, slightly off of their balance, and have been objects of
such commiseration to their dearest friends that the
insane asylum has been gravely hinted at as being a fit
place for men thus afflicted. The jealousy which exists in
all the well established professions will not tamely submit
to innovations which strike at what they conceive fundamental principles. They will not sit quietly by and see
prop after prop being knocked from under them, without
making a desperate effort to save intact the entire foundation as well as the superstructure. Loyalty to old established customs, methods and theories, impel them to inter-
pose the arm of defense. Entrenched as they are, behind
a breast-work of ignorance, they wage a conflict of error
against right, and although they may for a time succeed
in their assaults, truth being mighty, will ultimately pre-
vail.

No class of people are quicker to dispute the advance of
a thought or principle, which tends to interfere with what
they deem to be their rights, than the medical profession.
Medical ethics and medical etiquette are carried so far,
that they become ridiculous in the extreme. That species
of snobbery which will not permit one school of physicians
to practice or consult with another, is worse than ridiculous, and springs from pretensions as great as those which
characterized the Pharisees of old. This same spirit is
smaller, uglier, and more ungenerous than close communion, and would exclude all from the enjoyment of the

bright elysium beyond, who are not as holy as they. But
they stand upon a more lofty, but not less ridiculous emi-
nence than this. They profess to have a holy horror of the
practice of advertising. It is undignified, and weakens the
morale of the profession. It antagonizes the rule of their
enforced ethics, and some of these straight-jackets are
such sticklers for this all round foolish whim, that they
outwardly (but inwardly?) object to having even a modest
card, bearing simply their name and number of office,
inserted in a newspaper! And yet these M. D.s' names
may be seen in letters of gilt or gold, hanging upon a sign
above the entrance of their offices, or firmly nailed upon the
door: What is the difference between seeing a man's name
on his door, and reading it in a paper? What is the difference between seeing a man's name in the directory, and
reading it in a newspaper? This old, moss-grown idea
handicaps the physician of advanced thought and skill,
and keeps him from coming to the front, long years before
he could get there with both feet, if his ability and deserts
were made known through the ordinary and intelligent
channels—the wide-awake newspapers of the day. As long
as the M. D.'s run in the old ruts, they will continue in the
old ruts, where they have been from time immemorial.

A little more than a quarter of a century has passed since
the then small, quiet village of Kirksville, Missouri, received an accession to its inhabitants, in the person of a
man, who was at the time a pioneer in scientific discovery.
He boldly proclaimed the startling theory, that drugs were
not necessary to the life and happiness of the human race.
Like Robert Fulton, and others whom I have mentioned,
he was greeted with incredulous sneers, and at once became
the target for all sorts of idle and random shafts. He
was regarded as a mild fanatic by the people with whom
he had cast his lot, but faithful to his convictions, he con-

tinued to delve deeper and deeper into the mystery of science. Established theories had been exploded by his keen, logical researches, in his efforts to arrive at the hidden truth, and he became convinced that the Almighty Father of us all never decreed that the human stomach should be used as a chemical laboratory. Like the great Harvey, the discoverer of the circulation of the blood, he studied nature and the laws which govern and control the human organism, and after long years of ceaseless toil, filled with "hopes deferred" and bitter disappointments, he finally and firmly established the science of Osteopathy, the art of healing without drugs.

Dr. Andrew Taylor Still has lived to see the theory for which he long and patiently battled one of the fixtures in science, and is to-day reaping the fruits of the seeds which he has so intelligently sown. To such men as he, civilization owes its advanced position in art, science and invention. Their courage, persistency and lucidity of conception have been the levers which have lifted the world out of its ruts, and given it a new impetus in its onward strides.

Such men as he have been the benefactors of the race, for by their genius they have increased the aggregate of human comforts and human happiness.

Dr. Still, the discoverer of the basic principles of the new science of Osteopathy, founded the first school that taught the art of healing without drugs, and gave it the name of The American School of Osteopathy. He had been a regular physician from his twenty-first year until 1874, when his attention was seriously given to Osteopathy, or the new system of healing, which he afterwards developed and brought into recognition. His many years experience and practice in allopathy impressed him more with the thought that the practice of giving drugs in discase was a mistake, and his own researches convinced him there was a better system, a system which could be reasoned out from a deeper study of nature. To take up this line of thought, however, meant not only an abandonment of the profession to which he had given the best years of his life, but it meant, to a certain extent, the loss of reputation, and the still greater sacrifice of business. But what was the loss of standing in the old school of allopathy, and the sacrifice of property, if it were suffered in the cause of truth — in the achievement of something that would ameliorate the condition of the human race? Truth was more to him than all else besides, and having the courage of his convictions, he waged a struggle of more than twenty years. During this time he was subjected to many things. Many of his friends not only deserted him, but the most ignorant and prejudiced among them joined in with those who were the loudest in their jeers and scoffs. This treatment made him more determined to succeed in what he had undertaken. He had a message to communicate to man, and no opposition coming from man could prevent him from faithfully discharging his mission.

He began to test his theory among the poor of the community, and occasionally went from place to place to find a patient that would permit him to try his experiment. He continued his labors with indefatigable zeal, treating general diseases without medicines, until the results became so remarkable that many who were afflicted came to his home where they were compelled to remain

As the story has been carefully and truthfully told by another, I will here reproduce what has been said:

"About the year 1887, Dr. Still's practice had grown to such proportion that he found it impossible to attend to it alone. He then began teaching his new system to his son Harry. This experiment was so successful that his sons

Charles and Herman soon followed, with the younger
brother Fred, and a few intimate friends of his family.
By this means it was soon demonstrated that Dr. Still's·
new method could be imparted to others, for the sons readily became experts, and secured results in practice that
were considered almost as wonderful as had been the work
of the father. About this time the name 'Osteopathy'
was coined by Dr. Still and applied to his science.

"This first little private class conducted by Dr. Still at
his home slowly increased until about the year 1892, when
a charter for a school was taken out under the laws of the
state of Missouri, and Dr. William Smith, of Edinburgh,
Scotland, the present demonstrator of anatomy, was engaged to teach that branch. This was really the first class
in the school, and was looked upon as an experiment. Now
that Dr. Still finally understood the great truths for which
he had labored a lifetime, he was confronted with the
gravest problem yet encountered: 'How can Osteopathy
be taught to others that the work may be given to the
world?' Experiments in methods of teaching were now
necessary, and these extended over several years with vary-
ing success and disappointment until October, 1894, when
a new charter was granted, there having been some grave
errors regarding the power conferred by the first instrument. The school and infirmary were then conducted in
a little one-story frame building that stood where the new
building now stands. There was only one class, and really
the school was looked upon as a very small part of the work.
For several years the number of patients coming to Dr.
Still for treatment had been increasing rapidly until he
and his assistants had all the work they could possibly do.
At first patients came only from neighboring towns and
counties, then from adjoining states. In January, 1895, a
three-story brick building, fitted with all modern conven-

iences, was completed at a cost of $30,000, every dollar of which was paid with money earned in the practice of Osteopathy. The number of patients, which had about doubled each year, had so greatly increased that in the year 1895 over 30,000 treatments were given to sufferers from nearly every state in the Union.

"In October, 1895, a class of twenty-seven was enrolled, followed by a class of twenty-three in January, 1896. Of these two classes twenty-eight were from the state of Missouri, while the others represented five different states. These classes recited to one teacher in one class-room 20x25, the recitations taking up only two hours a day. In May, 1896, work was begun on an addition that doubled the capacity of the building, but even before that was completed, the rapidly increasing demands made it necessary to begin work on a second addition, which trebled the size of the original edifice. The whole building, which was completed in January, 1897, is four stories high, contains sixty-seven rooms, aggregating 30,000 square feet of floor space, and costing $80,000.

"The largely increased facilities were provided none too soon, for while the infirmary business has grown steadily, increasing at the rate of about 100 per cent. each year, the increase in the school attendance has been phenomenal during the last year, increasing nearly 500 per cent. in twelve months. While one year ago there were fifty students from six different states, reciting to one teacher in one room, there are now two hundred and eighty-three students, representing twenty-four different states and two Canadian provinces, using nine large class rooms, with lectures and recitations occupying the entire day, from 8 o'clock in the morning until 5 o'clock in the evening, with demonstrations held at least two nights each week."

OSTEOPATHY AS A SCIENCE.

"Osteopathy may be formally defined as the science which consists of such exact, exhaustive, and verifiable knowledge of the structure and functions of the human mechanism, anatomical, physiological, and psychological, including the chemistry and psycho-physics of its known elements, as has made discoverable certain organic laws and remedial resources within the body itself, by which nature under the scientific treatment peculiar to Osteopathic practice, apart from all ordinary methods of extraneous, artificial, or medical stimulation, and in harmonious accord with its own mechanical principles, molecular activities, and metabolic processes, may recover from misplacements, disorganizations, derangements, and consequent disease, and regain its normal equilibrium of form and function in health and strength."

To use a common-sense definition, the term "science," means a truth found out. In other words, it is a truth discovered. Man has never invented a single truth. He has only discovered it—developed the process by which it is ascertained. Truth has existed always, and man's investigations along any line of thought only bring to light the hidden law or forces by which it is made visible and operative.

It has its seat in the heart of God and embraces in its empire the entire universe, so that all truths manifest God, and truth is God. Truth is as eternal as our own rock-ribbed mountains. The storms of error may beat about it, and the fogs and mists and snows of centuries may envelop and bury it from mortal ken, but like the gods from the crucible it will emerge brighter and stronger.

Truth is absolutely immaculate and indestructible, and is the most omnipotent force in the universe. It is genuine, while falsehood is but the imitation. Imitations or counterfeits presuppose the existence of a higher ideal. Men simulate the truth to carry out their nefarious plans and purposes. We see this in every field and department of labor, and in all spurious inventions. Enter the realms of fiction! What is it that holds and fascinates the reader and furnishes food for thought and pleasant reflection? It is the semblance of truth the story bears. Sir Walter Scott blended with his matchless romance the truths of history. Through the warp and woof of Dickens' stories runs the shuttle of truth. He impaled upon the point of his magic pen the crimes and oppressions which disgraced some of the gently nursed and most tenderly caressed institutions of England. The thoughts which Shakespeare hurled upon the world with the force of ponderous projectiles will never cease to vibrate, because of the pungent truths which they contain. Take any of the great principles in science or philosophy and they live because of the truths they express.

Wordsworth says, truth is

"A staff rejected."

It is rejected until it has forced its way into the minds and consciences of men.

Nature herself has constituted truth as the Supreme Deity which is to be adored by mankind. She insinuates herself into the minds of men; sometimes exerting herself immediately, and sometimes lying hid in darkness for a length of time; but at last struggles through it, and appears triumphant over falsehood. Each truth sparkles with a light of its own, yet it always reflects some light upon another. The first truth colors all that succeed it, and each particular truth in its turn resembles a great river that divides into an infinite number of rivulets. Truth is the strength, wisdom, power, and majesty of all ages. Science being .hen a truth, each additional truth is added knowledge. Osteopathy invites the most scientific and scholarly investigation; being a new science, and its principles being utilized for the good of humanity, it is but right that its claims should be thoroughly understood, and that the inquirer should be fully informed as to its scope and intent.

The reputation it has already achieved as a science whose distinctive feature is the art of healing without drugs, was not born of any finely spun or mysterious theory, which required the comprehension of some recondite mind to understand, but it has been the result of facts, which are patent to the good common sense of every unprejudiced man or woman. Its culture and development have been of slow growth, but the plant has been carefully watered and nourished by its friends and expounders, until from its trunk are now shooting out the branches which will continue to spread and grow in strength until all suffering humanity may find healing in the grateful shade which they are destined to cast.

Osteopathy, reduced to its last analysis, embodies matter, motion and mind, which make up the trio of its working elements. "These indicate the comprehensive sweep of its radicals and its relations. These three terms suggest its logical unity and its chronological development, its analytical insight and its synthetic foresight. These reveal the organizing potencies of its initial premises, and the completed summary of its concluding propositions."

Scientific research in whatever direction it may be followed begins with the lowest substratum, which is matter. This forms the basic principle upon which are builded all scientific superstructures, and without these there could be no reasoning from cause, and no logical effects could be arrived at. Hence, in discussing this new science we should have a clear insight into the importance of living matter and of that vital force with which it is instinct. The abnormal conditions of the body always indicate the existence of disease, for no disease can be manifested where the functions of the body are perfect and are performing their proper work. Living and lifeless matter, although totally distinct, are often closely allied, for the lifeless matter is transformed into living substance when subjected to the chemical processes of the stomach. Food to the human organism is necessary to sustain life; but after it has been operated upon by the gastric and other juices which belong to the body it is constantly passing out as so much waste; a certain proportion of it remains, being transmuted into a living substance, where it eventually dies or becomes effete and is finally cast out. A writer has compared the living body, not inaptly to a whirlpool, into which matter is constantly entering. This, however, does not change the form or the individuality of the whirlpool in an external sense, but interiorly there may be changes. The body is the receptable for different kinds of food or lifeless matter, which work no perceptible outward changes, but these substances are expelled as new combinations entirely different to what they were when taken into the body.

Osteopathy asserts that all life in matter is a form of motion, which position is deduced from the principles of universal science. From this deduction we arrive at a more complete understanding of matter and motion, and a careful consideration of these two forces will enable us to more fully comprehend what osteopathy really is, and the foundation upon which the superstructure has been reared. Living matter analyzed shows that it is made up of a number of chemicals. It has been supposed that not less than six of these chemical elements unite in forming life. How many more is not known. The number, however, will be definitely ascertained as our knowledge in science increases. That may be in a few years, or it may require half a century, for the life germs or what they are is one of the most subtle and profound subjects with which the finite mind has ever had anything to do. The material basis of animal as well as of vegetable life is supposed to consist of proteids. These are one of a class of amorphous nitrogenous principles containing generally a small amount of sulphur, an albuminoid, as blood fibrin, casein, milk, etc. Each of these six elements has a force or an influence of its own. The oxygen gives vigor; hydrogen molecular mobility; carbon, sulphur, and phosphorus, allotropic properties. Herbert Spencer says: "All these peculiarities may be shown to be of significance when considered as attributes of living matter."

Proteids then form the important "substratum of the human organism, and through their agency the chemical and mechanical processes of the body are effected."

Taking it for granted that each human body is furnished by nature with a chemical laboratory, in which the same kind of changes are carried on as they are produced by the working chemist in his shop, the conclusion is inevitable that these proteids and the substances with which they are assimilated and allied are manufactured in the human organism. If these substances were not in a constant process of manufacture and reproduction life would become extinct by wasting. The natural heat of the bodily tissues if not supplied or reproduced by the chemicals of the body itself would lose its energy and life-giving principle, and the living matter would become lifeless matter. Many things in nature, externally considered, grow by accretion, while internally considered, or in the human body, they grow by absorption or interpenetration. The juices and liquids permeate every part of the system through the numberless ducts and channels, and the substances thus borne to their proper destination fulfill their mysterious mission, and in the course of time are expelled from the body. This same process is repeated as long as the body remains a living organism. The recuperative powers of the body are most wonderful, and osteopathy claims that these forces are so great and self-sustaining that even after they are deranged by disease or disorganized from external contact with outside influences they can be readjusted and made to pursue the even tenor of their way without the aid of drugs.

"We may perceive how extraordinary these powers and properties of the body are, unaided by medicine, by supposing a locomotive engine to possess like powers to carry on a process of self-repair, in order to compensate for wear; to grow and increase in size, detaching from itself at intervals pieces of brass or iron endowed with the power of growing up step by step into other locomotives capable of running themselves, and of reproducing new locomotives in their turn. Precisely these things are done by every living body, constantly and for years, by the mechanical and molecular activities of matter and motion."

The body is not only a piece of mechanism, but is a machine, infinitely more complicated than anything that could possibly be devised by man. The physiological mechanism is self-asserting, self-propelling. The locomotive machine is not. It runs at the will of the mechanic or engineer who starts it, and it stops at his will or pleasure. The human machine needs no appliances or lubricants from without to continue its movements. When it was constructed these were made with it. It makes no differ-

ence whether the mind (the engineer) is present or absent;
every part of the machine performs its functions.

The locomotive is made with reference to its durability
and strength; the time of its useful existence is almost accurately known, and its strength is also known. The
human organism upon the contrary is as yet only partially
understood. If free from disease and other causes which
hinder its operation it will run for an indefinite period of
time, in the meantime furnishing its own propelling forces.
These forces constitute the oxygen, which is the dominant
essence of the body, and without which it could neither
grow nor maintain health. This chemical attribute is both
destructive and healthful. It destroys the impaired part
of living tissues, and builds up those that are good and
healthful. The chemist can make the best of oxygen, but
it is of no force without being supplemented by nature.
It must be supplied by nature, else it cannot produce the
results that are required by the system. The working laboratory of the scientist cannot build it. The well being of
the body demands the pure oxygen of nature and cannot
exist in a healthful condition without it. The blood becomes stagnated, and the muscles become tired, the heart
moves sluggishly, the impulse to digestion disappears, food
passes through the stomach unused, the blood does not
circulate through the brain, and the head begins to ache.
The dead tissues are not carried off by exhalation, but are
collected in different portions of the body, where they produce and develop disease.

"Oxygen is the first, foremost, greatest, and most active
element that can be taken into the system; no other matter
can equal it in importance, whether it is found in what we
eat, drink or breathe. A person is nearly three-fourths
oxygen. In addition to oxygen there are thirteen other
elements that necessarily enter into the materials of the

body, and before they can enter into the human organization they must have been organized by nature and thus be
charged with the power of becoming a part of life.

"But this organization must take place in some vegetable.
Man eats two kinds of food: first, that of food eating creation; second, vegetation. The first we call meat; the second is selected from the vast division of life which in-
cludes plants, roots, herbs, seeds, grasses, fruits, etc. It
is a fact that all unorganized matter is unfit as a nutriment
for the human body. The absurdity, for example, of taking iron in any disorganized form to supply the lack of this
element in the blood is seen in many cases of invalids who
have suffered from medicines which furnished this material. There is no medicine or mineral now on the market,
or possible to be made, which can furnish iron to the body
in an organized form. Thousands of people of feeble
constitutions are periodically eating pills or taking syrups
or other mixtures containing iron, and vainly imagine that
it may thus be restored to the blood. The famous French
physician, J. Francis Churchill, quotes from Trousseau as
follows:

" 'Iron hastens the development of tubercles. The iron
may induce a fictitious return to health; the physician may
flatter himself that he has succeeded; but to his surprise
he will find the patient soon after fall into a phthisical
state, from which there is no return.' "

We spoke of the elements which are necessary to sustain
life. These are oxygen, phosphorus, carbon, hydrogen, ni-
trogen, calcium, sulphur, sodium, chlorine, flourine, iron,
potassium, magnesium, silicon. In the human body there
are seventeen combinations of these fourteen elements of
food material: Water, gelatin, fat, phosphate of lime, al-
bumen, carbonate of lime, fibrin, flouride of calcium, phosphate of soda, phosphate of potash, phosphate of magnesia,

6

chloride of sodium, sulphate of soda, carbonate of soda,
sulphate of potash, peroxide or iron, and silica.

These elements are supplied by food, and are, day by day,
some of them, taken into the system. If any are absent
sickness or disease will be the result.

We cannot quite agree with some writers who assert that
mind governs all the molecular combinations of the human
body. We intimated in a preceding page that the mechanism of the body moved along, so to speak, regardless of
the mind. This is demonstrated in many ways. The machinery operates just the same when we are in a natural
sleep. It operates just the same when we are under the
influence of an anesthetic—when the self-assertive power
of the mind lies entirely dormant—so much so that the
excision of a limb is not felt by the brain. It operates
just the same in an idiot who is recognized as possessing
no mind at all. Of course the atomic constituents of the
human organism make the body as a grand whole, and volition which emanates from the mind, exercises a mastery
over the machine or any one of its parts or sections, and
endows these parts or sections with the power to manipulate or execute. The hand is employed to execute the purpose of the mind, but the functions of the body and the
complicated mechanism which characterizes it may be just
as vital and as full of energy if the mind were a blank.
"As a man thinketh so is he." This has reference to his
habits and practices in life—to the character which he
bears, and which are the result of his thinking machine.
This thinking, however, does not change the mechanism of
the body, nor its appearance. Some idiots have as finely
developed physiques as can be seen, and yet they have
neither thinking nor reasoning faculties.

Osteopathy goes back to first causes or principles and
recognizes God as the beginning, or God as mind. In

other words: that God, mind, or spirit, whatever we may
choose to term it, is all and in all, and that in man is manifested the highest type of creation. Man, we are told, was
made in God's image, which, of course, means in a spiritual
sense—in the sense of mind. Man then, as an intellectual
being, reflects the likeness of the great first cause, which is
God. God is an omnipresent, illuminated spirit, and man
is a part of God, reflecting as he does his image. God is
in all and over all. "In him we live and move and have
our being." We can never get outside of God, which is the
good, or the boundless love. The foundation of all science is laid in the thought that nature is the perfection of
intelligence. Nature's plans and provisions are adequate
to the promotion of vigorous health, when let alone and
not interfered with by man, who, we are told, has "sought
out many inventions." Dr. Still says: "That a natural
flow of blood is health; and that disease is the effect of a
local or general disturbance of blood." No intelligent
physiologist will dispute this proposition, for the blood
which circulates in the principal vascular system not only
carries nourishment to every part of the body, but it brings
away the waste products which are expelled. Any thing,
therefore, which disturbs or hinders the natural flow of
blood through the arteries, capillaries and veins, trenches
upon health to a certain extent. If one of these arteries
become clogged or impaired in proportion to its importance
to other arteries of the vascular system, so will it affect
general health. A man may lose an arm and yet live just
as long and do as much work in certain fields of labor as
the man with two good arms, but for all purposes he cannot successfully cope with the man with two arms. The
blood may be impaired in its free circulation at some important point. If it is, the general health may be maintained, but not at its full vigor. A single broken cord,

though its effects may not be noticed among a thousand other perfect cords, yet creates a certain amount of inharmony or discord, which exists whether it be detected or not.

The body must be continued in a normal condition, and this can be done only by keeping intact all the functions, however minute they may be, which enter into its construction as essential and constituent elements. The laws of its organization, both in health and disease and its adjustment by the mechanical appliances of the body, constitute the groundwork of Osteopathy and show that its teachings are in harmony with the most advanced scientific thinkers of Europe and America and their application in the numerous branches of the medical practice of to-day.

Life and death, health and sickness, are problems which can be solved only by an investigation into the vital forces of the body, upon which it is builded and perpetuated. This problem is the one which Osteopathy has undertaken to explain, and which it is explaining in accordance with the laws which control and regulate the human organism. It is doing so in a scientific way, and along the only line that nature suggests.

Dr. Still says that "Scientific diagnosis cannot be based alone upon symptoms or histological findings; it must be based upon ascertainable and demonstrable facts." The composition of the vital substances can only ultimately, he says, be determined by the microscope, aided by whatever mechanical abnormalties a profound knowledge of anatomy will detect. This intimate knowledge of anatomy, based upon evident mechanical molecular conditions, will enable the skilled osteopath to make the proper diagnosis. The medical practitioners, owing to the unscientific character of the diagnosis they make, know no more after they have made it than they did before. This is attributable to their ignorance of the formation of the diseased cells. The average physician knows but little about the matter which produces these cells. His training has not led him to consider their character and composition. He relies more upon what has been the common practice, or upon what has been laid down by his contemporaries, than he does upon his own clear and comprehensive insight. In other words, his diagnosis is the result of what he believes to be certain symptoms, and these symptoms indicate certain diseases. His knowledge of the profession which he practices will not permit him to go into the causes which lie at the base of the disease which the symptoms indicate. He is satisfied to proceed along the old rut, and prescribes accordingly, the thought not having once entered his mind that he should know more of the disease than its mere symptomatic appearance indicates—that is to say, he is satisfied when the symptoms point the existence of any certain disease, the name of which is given in medical books. He goes no further. He does not dare to extend his explorations beyond the limit which has been laid down. His superficial knowledge of anatomy and physiology, and the deeper causes from which diseases spring will not warrant him in invading the field of causes upon his own volition.

For pains or cramps the patient is given narcotics, little thinking or caring that these drugs pollute the blood and thus influence the function of the nerves. They may lull the pain or ease the cramps, but at the same time they may so affect the sympathetic system that its natural work may be retarded, and the vital forces lowered; or they may cease to perform their functions altogether.

The ordinary doctor prescribes the most noxious and even poisonous drugs as medicines. These number many hundreds, and each has its effect upon the system; not in

ultimately building it up, but in reducing its normal vitality. He tries remedies as experiments. If one does not
produce what he conceives to be the desired effects he
tries another, and continues these experiments until he
has exhausted the whole category of remedies without obtaining what he wants in the condition of the patient.

Dr. Radcliff said a short time before his death: "Since
I have grown old in the art of healing, I know more than
twenty diseases for which I have not even a remedy."

The modern school of medicine makes the mistake in
not understanding the fact that the living matter from
which the organized cell is produced must first become deranged or impaired. "The cause of the derangement is the
first question to be settled, and this cannot be decided upon
a scientific basis without a thorough knowledge of the condition, composition, and control of the living matter of the
body, together with its mechanical relations and molecular
activities. If it were possible to treat diseased cells, or
parasitic bacteria directly by chemical means, osteopathy
asserts that it would not avail anything, for the reason that
the same conditions which produced the diseased cells
would construct others in the place of those destroyed,
which is evidenced in cancer cells which, if removed, are
succeeded by others.

"As a logical conclusion, it will be seen that healing by
such methods would be impossible without creating first
a normal condition of the living matter of the body, and
removing all obstructions to the natural flow of the blood.
The nerves which impart health and strength depend upon
the blood, so that after all everything in reference to life
and health depends upon the natural flow of the blood.
Upon this question of blood, and its importance to the continued vitality of the system, osteopathy occupies an advanced position. It has penetrated the secret recesses of

the body, and studied the chemical constituents of its materials of supply in air and food; it has studied the character and quality of the living matter and vital substances,
the laws of their chemical construction, the cells, tissues,
organs, etc., and is convinced beyond all question of a doubt
that no laboratory of the chemist or lotion of the pharmacist can approximate the formulations of nature. Osteopathy has demonstrated that the resources and remedies
which are in the body, and given to it by nature, are sufficient to restore it from disease to health without drugs or
artificial remedies."

Dr. Still says: "The brain of man is God's drug store,"
which is proven by the fact that it possesses all the char-
acters of a machine which have been made for that purpose. Its mechanism, so elaborate and wonderful, manu-
factures and manages every chemical necessary to every
function of the body. The brain constitutes the laboratory
of life, and this laboratory is placed by Osteopathy in the
body. Dr. Still further says: "That here," meaning the
brain, "and here alone are superintended and supplied the
processes and products in the exact quantity and quality
which the body needs, by which the vital functions of digestion, absorption, assimilation, growth and health are
maintained. Here, also, are conducted the remedial processes by which the body recovers from sickness and disease,
in the use of the proper materials furnished to the body,
and by means of the mechanical appliances possessed by the
body for this express purpose. The powers of the body
are such that it can bring together, in mouth, stomach, and
intestines, with the assistance of the liver, gall-bladder,
pancreas, spleen, and the entire circulatory, secretory and
excretory systems, the materials for its subsistence, in such
close contact, and under such wonderful conditions of heat
and solution can infuse their elements with such affinities,

and make those affinities so operative, can exert such influences that forthwith some new substance is wrought into its own being with powers and energies the most subtle or the most tremendous. It may be death to any and everything inimical to the body, or it may exercise on the organism the most blessed virtue, restoring the wasted tissues, reanimating the vital flame, and carrying into the most secret recesses of life the sweet influences of health. This is something of the birth-power of the brain, of the mastery of the mind."

The scientific assumption that chemicals can be made by man similar to those which nature has furnished the body the power to manufacture, and similar to those possessed by the body is untrue. There is an absolute difference between the chemicals of the body and those prepared by man. The chemicals of the druggist or chemist are only approximations or imitations. They fall far short of those that are wrought by nature, as medicine is at best only inorganic, and only from one-tenth to one per cent of the matter of the brain is inorganic. Drugs ultimately destroy the irritability of the nerves and take away their functionary power. The druggist obtains artificial results only, while the chemicals produced by the human organism are natural and genuine.

Unnatural combinations prepared for the system as medicines destroy the vital energy of the nervous structures and can be of no possible efficiency as a nutriment. Only organized matter becomes assimilated in the body, and this work is done by nature. The storage battery of life and health are the brain and nervous system, and these become disorganized and weakened when subjected to the influence of drugs. The intelligent osteopath who studies the nerves, their origin, distribution and chemical constituents understands more of nature and the laws by which nature is governed than the man who seeks to cure ailments by drugs. Osteopathy teaches that the constant efforts of nature are towards a restoration of health, and that all functional disorders can be removed and healed by a perfect knowledge of anatomy and physiology without the use or application of drugs in any form whatever. Having a trained and sensitive touch and a thorough knowledge of the nerve centers in general, and of their exact location, the osteopath possesses the means of healing at his finger's ends. The following illustration is one to the point. "As the violinist knows what notes to touch, and easily and intelligently slips his fingers along the strings and gets such tones and tension as produce rhythmical harmony, so an osteopath has profoundly studied the human organism, with all the aids of literary research, morbid anatomy and normal life in all its delicate and dexterous forms and forces and health-giving functions, and by skillful operation secures the natural equilibrium and healthful activities of the human frame."

In reference to the word "osteopathy," Dr. Still says it more correctly describes the science than any other word that might be chosen. The word embodies one of the great ideas of the science. The bony framework of the body is that part upon which the true order of the body depends. The bones are the most substantial underlying landmarks of the body. They constitute the hard, unyielding substratum upon which all other structures are built, and upon which they depend for permanence of position and location. The bones constitute the foundation of the bodily superstructure. Besides, they are the fixed points from which the trained anatomist may correctly explore for disorder in the mechanism. The body is an embodiment of all the principles of. mechanics, of physics, of hydraulics, all architecture, and all machinery of every kind. There are nearly four hundred mechanical principles that have their finest practical illustration in the human body.

"Here are found all the bars, levers, joints, pulleys, pumps, pipes, wheels, and axles, ball-bearing movements, beams, girders, trusses, buffers, arches, columns, cables, and supporters, known to the most advanced mechanical science. These constitute its anatomical mechanics, which require the minutest study and mastery by the osteopathic student and operator. Then, there are the principles and philosophy of electricity, magnetism, of fluids at rest and in motion—hydrostatics and hydrodynamics—capillarity, diffusion of liquids, and osmosis, and their manifold application to circulation, absorption and secretion. Then there are pneumatics, or the physics of gases, and their application in respiration. There are optics, the action of prisms and lenses, the mechanism of light, refraction, polarization and the interference of light. There is sound as related to sympathetic vibration and resonance; and heat, in its conduction, convection, and radiation, as related to the body. There is also dynamics, as operative in the mechanics of matter, force and gravity, in the body. These constitute its physiological physics, which must be considered in mastering the forces and motions of the body. And all these possibilities of mechanics and physics are related to the bony framework of the body. The bones, then, are pre-eminently the means by which the physics and dynamics of the body are made operative and effective.

"Concerning surgery, the doctor says there is a necessary place for it in some emergencies of osteopathic practice, there are many abnormal conditions of the body that require a proper surgery, and osteopathy is training its operators in the accessory science of surgery, and is pursuing a wise physiological course between the harmful use of anæsthetics, that are almost invariably used in such operations, and the neglect of the health of the body under abnormal conditions that really require the services of a skilled surgeon. Osteopathy thus makes a new advance upon the science of modern surgery, and its course has been vindicated by the success of its methods along this line."

MUTTERINGS OF DISCONTENT.

The signs of the times point unerringly to the fact that
drugs and medicines are losing caste among men of the
highest culture, in the old world, as well as in the new.
And this condition of affairs is steadily growing worse, so
far as the old school of medicine is concerned. Men of
thought, both out of and in the practice of medicine, are
becoming more and more impressed with the idea that
there is but little that is true or genuine in the practice,
and much that is in the business of the druggist and pharmacist that descends to the level of trickery and delusion.
The science of healing, known as Osteopathy, has its foundation in facts that are as plain as the first letters of the
alphabet; it teaches that the laws of nature are the only
true and reliable guides in the treatment of disease, and
that a departure therefrom leads us into by-paths that are
dark and devious and brings us into contact with forces
we do not understand, and from which, if we rightly understood them, we would flee with disgust if not with right-
eous indignation. In proof of the fact that drugs and
medicines are waning in their influence among many of
the ablest professional men in the country we subjoin a
number of testimonials that will form a most interesting
chapter to doctors who cannot see any virtue in osteopathy.
It will show to them that there are two sides to this great
question which is gradually arresting the attention of thinking men the world over. These mutterings are but the pre-
monitory signs which indicate the coming of the great
upheaval which is destined to overturn some of the most
firmly rooted, rock-ribbed theories which have been held

to for more than two thousand years by those who have followed in the footsteps of Esculapius and Hippocrites:

(As tabulated by the Journal of Osteopathy:)

John Mason Good, M. D., F. R. S.: "The science of medicine is a barbarous jargon. My experience with Materia Medica has proved it the baseless fabric of a dream; its theory pernicious. The effects of medicine on the human system are in the highest degree uncertain, except, indeed, that they have destroyed more lives than war, pestilence and famine combined."

Dr. Evans, Fellow of the Royal College, London: "The popular medical system has neither philosophy nor common sense to commend it to confidence."

Prof. Valentine Mott, the great surgeon: "Of all sciences, medicine is the most uncertain."

Sir Ashley Cooper, the famous English surgeon: "The science of medicine is founded on conjecture."

Prof. Gregory, Edinburgh Medical College: "Ninety-nine out of every one hundred medical facts are medical lies, and medical doctrines are, for the most part, stark, staring nonsense."

Dr. Cogswell, Boston: "It is my firm belief that the prevailing mode of practice is productive of vastly more evil than good, and, were it absolutely abolished, mankind would be infinitely the gainer."

Dr. Marshall Hall, F. R. S.: "Thousands are annually slaughtered in the sickroom."

Sir John Forbes, Fellow of the Royal College of Physicians: "No systematic or theoretical classification of diseases or therapeutic agents ever yet promulgated is true, or anything like truth, and none can be adopted as a safe guidance in practice."

Bostwick's History of Medicine: "Every dose of medicine is a blind experiment upon the vitality of the patient."

Prof. B. F. Parker, New York Medical College: "The drugs which are administered for scarlet fever kill far more patients than disease does."

Prof. E. R. Peasley, M. D., New York Medical College: "The administration of powerful medicine is the most fruitful cause of derangement of the digestion."

Prof. Alonzo Clark, New York College of Physicians and Surgeons: "All our curative agents are poisons, and, as a consequence, every dose diminishes the patient's vitality."

Dr. Oliver Wendell Holmes: "Mankind has been drugged to death, and the world would be better off if the contents of every apothecary shop were emptied into the sea, though the consequences to the fishes would be lamentable. The disgrace of medicine has been that colossal system of self-deception, in obedience to which, mines have been emptied of their cankering minerals, the entrails of animals taxed for their impurities, the poison bags of reptiles drained of their venom, and all the inconceivable abominations thus obtained thrust down the throats of human beings from some fault of organization, nourishment or vital stimulation."

Prof. Geo. B. Wood, M. D., University of Pennsylvania: "We have not yet learned the essential nature of the healthy actions, and cannot, therefore, understand their derangements."

Prof. Magendie, the distinguished physician of Paris: "I hesitate not to declare, no matter how surely I shall wound our vanity, that so gross is our ignorance of the real nature of the physiological disorders called disease, that it would, perhaps, be better to do nothing and resign the complaint we are called upon to treat to the resources of nature, than to act as we are frequently called upon to do, without knowing the why and the wherefore. of our conduct, and its obvious risk of hastening the end of the patient."

Dr. Talmage, F. R. C.: "I fearlessly assert that in most cases our patients would be safer without a physician than with one."

Joseph M. Smith, M. D., College of Physicians and Surgeons: "All medicines which enter the circulation poison the blood in the same manner as do the poisons that produce disease."

Dr. Broady, "Medical Practice Without Poisons": "The single, uncombined, different and confessed poisons in daily use by the dominant school of medicine number one hundred and seven. Among these are phosporus, strychnine, mercury, opium and arsenic. The various combinations of these five violent poisons number, respectively, twenty-seven combinations of phosphorus, five of strychnine, forty-seven of mercury, twenty-five of opium, and fourteen of arsenic. The poisons that are more or less often used number many hundreds."

Prof. N. Chapman, "Therapeutics and Materia Medica": "One half of all who are born die before they reach seventeen years. One half of all born in our cities die before they reach three years of age. The average man, according to statistics, does not live out half his days. The responsibility of the medical system for this sad uncertainty of human life cannot be questioned."

Dr. Raymond, the eminent physiologist: "In regard to skepticism in medicine, unfortunately it was the doctors who set the bad examples. It is said the practice of medicine is repulsive. I go further and say, that under certain conditions, it is not the practice of a reasonable man."

Sir John Forbes, M. D., F. R. S.: "Some patients get well with the aid of medicine, some without it, and still more in spite of it."

Prof. A. II. Stevens, College of Physicians and Surgeons: "The older physicians grow the more skeptical they become of the virtues of medicine, and the more they are disposed to trust to the powers of nature."

Prof. B. F. Parker, New York: "Hygiene is of far more value in the treatment of disease than drugs. As we place more confidence in nature and less in the preparations of the apothecary, mortality diminishes."

Prof. J. W. Carson: "We do not know whether our patients recover because we give medicine, or because nature cures them. Perhaps bread pills would cure as many as medicine."

Prof. Magendie: "Medicine is a great humbug. I know it is called a science. Science, indeed! It is nothing like science. Doctors are mere empirics when they are not charlatans. We are as ignorant as men can be. Who knows anything in the world about medicine? Gentlemen, you have done me the honor to come here and attend my lectures, and I must tell you frankly now, in the beginning, that I know nothing in the world about medicine, and I don't know anybody that knows anything about I repeat it, nobody knows anything about medicine. I repeat it to you, there is no such thing as medical science. Let me tell you, gentlemen, what I did when I was a head physician at Hotel Dieu. Some three or four thousand patients passed through my hands every year. I divided the patients into two classes. With one I followed the dispensatory and gave them the usual medicines without the least idea why or wherefore. To the other I gave bread pills and colored water, without, of course, letting them know anything about it. And, occasionally, gentlemen, I would create a third division, to whom I gave nothing whatever. These last would fret a good deal. They would feel they were neglected (sick people always feel they are

neglected, unless they are well drugged, the imbeciles!),
and they would irritate themselves until they got really
sick; but nature invariably came to their rescue, and all
the persons in the third class got well. There was a little
mortality among those who received but bread pills and
colored water, and the mortality was greatest among those
who were carefully drugged according to the dispensatory."

The above excerpts are "confirmations strong as proofs
of holy writ" of the theory which is upheld and sustained
by Osteopathy. These are not to be considered as mere
isolated passages thrown off at random in the course of
lectures or in conversation by weak or indifferent practitioners of medicine; they represent the most mature
thought as the culmination of years of practice and observation by some of the most learned men in the professional ranks; men who have graced the highest positions
in the royal medical colleges of Europe and England; men
who are at the head of their profession in America; men
who are authors, having written much and thought deeply
and long upon the so-called science of medicine. These
are a few confessions which they have honestly made, and,
in spite of everything that can be done to unfavorably
criticise their significant utterances and everything that
may be done to so torture and twist their plain meaning,
they add immeasureable strength to the position of the
Osteopath and are strong in their condemnation of drugs as
remedial agents.

Were these the utterances of the ordinary uncultured
"pill-doctor," who has lived all his life in an obscure country village, or upon a little frequented highway, then they
might be easily pohed away as unworthy of being repeated
to intelligent people, but when they come from the highest
sources, the acknowledged head and front of the medical
profession the civilized world over, they should be calmly

7

weighed and given the serious consideration that they demand. These men are not knowingly, either in belief or practice, Osteopaths; the majority of them, perhaps, have never heard of the new science; hence their testimony in its behalf is of infinite value to the honest inquirer who is seeking a better way in the art of healing than that which leads through the shops of the apothecary and the retorts of the chemist.

If the foregoing witnesses to the deleterious effects of drugs do not carry sufficient force to convince the most confirmed believer in the old theories advanced by Materia Medica that there is something wrong in the use of chemicals, and the practice of medicine as it has been followed for centuries, we will produce another class of witnesses in the shape of extracts from a few of the medical journals which are published in the United States. These are facts gleaned from publications which cover a period of only two months in the year 1897.

Post-Graduate: "This country is literally being flooded with circulars and preparations, synthetic and otherwise, from the enterprising manufactories of Germany, and the readiness with which their testimonials are accepted and their drugs dispensed, which are useless if not harmful, reflects little credit upon the average intelligence of the American physician. One hundred and seventeen new drugs were placed upon the market in Germany within the short period of six months. If experience teaches anything in this world, it is that we are in need of fewer drugs, and of considerably more common sense in the practice of our profession."

Cleveland Medical Gazette: "It is a melancholy fact that there is not always sufficient concert of action in dealing with great questions outside of, but affecting the medical profession. The fact is, the medical profession has never awakened to a sense of its own power, social, political. It talks now and again about reforms, and complains continually of this or that evil which ought to be removed. But its efforts and its objections to right the wrong are as the querulous petulance of puny childhood to the might of gladiatorial manhood, when compared with what might be done, if the profession earnestly aroused itself to make an effort."

Willard H. Morse, M. D., F. S. Sc., Samatological Chemist in July, '97, Medical Brief: "A letter came one morning from a widow. There was this question in it: 'What can I do? My boy eats opium.' I knew that young man three years ago, the ticket agent in our city. He had had pleurisy, and I prescribed morphine. It made an appetite, and he had taken the habit. The habit readily acquired his slavery. It is a habit of suicide and the ill custom of prescribing soothing syrup, laudanum and morphia is the fault. I know it by sad experience, by seeing that boy raise his hand against his mother; by seeing him behind the bars of a prison cell; by standing with that widow by my side, and Mrs. Browning's lines ringing in my ears:

> " 'And that dismal cry rose slowly,
>   And sank slowly through the air,
>   Full of spirit's melancholy,
>   And eternity's despair.' "

Dr. Cram, specialist in Practical Medicine, "Brief": "It is not only from a multitude of such cases about us that defy the physician's art-cases within the circle of our own observations—that sadden, shame or shock us, but it is more particularly from the great army of chronic cases throughout the country that we obtain proof of the inefficiency of medical practice, and proof that is glaringly 'barefaced.' That it is also destructive may be seen in the astounding fact proved by statistics that there are over twenty deaths in practical medicine to each one that occurs in surgery and obstetrics combined.

"Viewing the medical education of the past, as it has come to us from the medical professors, the text-books and medical journals, we must conclude that the essential nature of disease has never been taught in connection with any of the diseases of this department. Here is the chief reason for the heavy death rate—a profound ignorance of the essential nature of disease."

Editorial in July, '97, "Brief": "There are firms scattered over the country whose business it is to prepare substitutes in odor and taste for standard pharmaceutical preparations. These substitutes are not the same thing at all, because it is impossible to accurately analyze any organic mixture, and because they contain the cheapest grade of drugs and chemicals. Many chemicals are injured by age and atmospheric influence, and many dried roots and leaves are absolutely valueless in the preparation of medicines. Yet these are cheap, and it is these which are used in making piratical substitutes which have no reputation to sustain. (The doctor who prescribes opium and astringents in diarrhœa due to intestinal fermentation and putrefaction is doing his best for the undertaker. And yet there are no better remedies than those named in the indicated conditions.)

"There is a great deal of misdirected energy in the profession at this time. The time devoted to getting a medical education has been greatly extended. Medical students apparently enjoy greater advantages than their preceptors did. In reality they emerge from colleges far less fitted to cope with disease than the old-fashioned doctor, who served as prentice hand to the country doctor a year and then secured a diploma on two terms of the lectures crowded with primal truths and practical facts.

"Our present methods of medical education tend to make closet doctors rather than practitioners. Men do not go to medical colleges to become naturalists, or to study the biography of eminent physicians. They are chiefly interested in learning how to cure sick people. Too much stress cannot be laid upon anatomy and physiology; these should be thoroughly mastered; they form the ground-work for all the rest."

Editorial in Homeopathist Recorder, Aug., '97: "Tens of thousands of victims of heavy old school and proprietary medicine drugging could be cured by staying at home and substituting bread pills for the drug mixtures."

Dr. Pratt, in July Journal of Orificial Surgery: "Preju-dice is melting away, hostilities to innovations are becom-ing enfeebled; all hindrances to progress are being torn down; the tyranny of ignorance and conceit is being rapid-ly overcome; medical monopolies are passing away. Drugs and knives and local applications no longer constitute a complete medical armamentarium. The part which men-tal and emotional forces play, not only in the functional activity of all bodily organs, but also in pathological for-mations, is at last being recognized by medical leaders, and also to a considerable extent by the rank and file of the profession. The value of Osteopathy as a remedial agent will in due time be recognized."

Dr. W. N. Mundy, in July Eclectic Medical Journal: "The physician does not progress in this channel very rapidly, but the pharmacist is making remarkable strides in flooding us with new remedies, which for a season prom-ise everything and do nothing. The physician has ceased to prescribe; the manufacturing pharmacist saves him that trouble by preparing for him his heart tonic, kidney com-

pound, analgesic and laxative compounds, cough remedies
and a host of others. There is danger that he will cease
to think; he won't have to; the pharmacist will do it for
him. He will become a mere machine, automatically prescribing that which has been prepared for him.

"We are personally acquainted with physicians who stand
high in the estimation of their colleagues and the public,
who use this method to a very large extent, and even use
what is styled patents. We would hardly call this science
or skill. There is no penetrating or comprehensive knowledge displayed in such methods. No skill required or displayed. So long as such methods are pursued, therapeutics
can never make any progress, nor can there be any certainty
in it."

Editorial in the above: "It is undoubtedly true that as
many people now resort to the patent medicines as are
treated by physicians. Formerly, much harm was done by
this self-medication, and the doctor rubbed his sides and
laughed, knowing that many of the victims would pass
into his hands to be cured, not only of the original complaints, but in addition, of the aggravations brought on by
the injudicious self-medication. Now, however, in many
instances, the laugh is on the other side.

"Probably it is beyond estimation how many 'headache'
cures, sleep-producers and cough cures, are yearly sold over
the counter in the United States. The employment of the
many coal-tar introductions is a cause of great concern, if
not of alarm. The extent of impairment and destruction
of the nervous system that is sure to follow, and to tell
upon the future unborn cannot even be conjectured; but
come it surely will."

John Uri Lloyd, Ph. M., in Aug. Eclectic Medical Journal: "In no place do we find this helplessness of man in
the face of nature better depicted than in that of medicine.

In no other field so close to science does science stand
aside, while chance stalks on and on, scattering to the
winds the foresight of scholars and of those who hope to
calculate by rule.

"Regardless of the eulogistic opinions of those who,
wrapped in optimistic or egotistical mantles, hold aloft the
phantom banner on which is inscribed, 'The Science of
Medicine,' I believe there is no true science of medicine,
unless it be that chance and empiricism dominate the
progress of medicine.

"Empiricism rules medicine of to-day with an iron hand;
chance leads with irresistible force, and in the clutches of
these two mighty agencies the earnest men who work and
think, and cry aloud in their hearts and in their souls for
a law that will turn disorder into method, that will create
science out of chaos, are helpless. There are several sections in medicine, all alike struggling to excel; none are
perfect. It matters little whether I speak to him who
cries: 'We are the people,' or yet to him who says: 'We are
the people;' tis but the change of the accent of a word.
No man is crying the truth when he cries aloud—'We have
the science of medicine.' Until the leaders of his peculiar
section evidence that fact by laws that agree one with
another, and that their followers can grasp by methods
common to true science, his science (?) in the eyes of men
of science is empiricism.

"By no means known to man can you of the medical
profession, from a scientific standpoint, establish the therapeutic value of any of these drugs. Neither can you by any
means at your command, tell why they possess their peculiar qualities. Your professional work neither gives a law
for the first nor second part, and you have no rule to join
the first to the second. The engineer will project his line
and calculate its course with precision; the chemist will

predict the future compound by law; the botanist classifies new plants that come to his hand, by scientific relationships, but the physician struggles in experimentation, ignorant of any law that will tell whether a new vegetable drug will purge or vomit."

Editorial in July Medical Fortnightly: "With the continued mortality from chloroform and other narcosis, we are reminded that as yet medical science has not reached that point where it can prevent such acidents, nor successfully overcome them when they have occurred. Both agents are toxic. How foolhardy for an otherwise skilled and conservative operator to entrust this very important duty to a medical tyro not familiar with his physiology, much less with the physiological action of drugs, to say nothing of his inability to recognize the important signs of danger, and to know what to do if danger even is imminent."

July Modern Medical Science: "It is fruitless to stimulate cell proliferation by tonics, which cannot feed their progeny, nor find or produce nourishment for them. Neither can the tonics go on stimulating indefinitely. The power of being stimulated is diminished by every application of the unnatural stimulation, until it is at length annihilated. At the same time, and by the same unphysiological, or rather anti-physiological intrusion, the functions of digestion and nutrition are upset; and verily the last state of the patient is worse than the first, or than the first would naturally have become."

So reason our own medical journals. The Osteopath sees much in these inklings, indicating a change of sentiment, from darkness into light, from error into truth, and rejoices at every genuine change of base, when it is seen that the ground that has been so long occupied is wholly untenable.

In reality, there is no reason why the so-called medical science does not advance to a higher plane of thought; generously and wisely let go of the old dogmas, which ought to have long since been exploded, and stand where the dawn of the incoming new light will shine upon them. The epidermis of the rhinoceros is never so thick but that it may be penetrated by the sunlight, which that animal often seeks in the most torrid clime. Long conceived opinions lived and practiced are hard to throw off, even in the face of absolute proof of their complete worthlessness. A man, however, who lives at this, the close of the nineteenth century, should make at least one effort to put himself in harmony with the spirit of progress that seems to be advancing along every line of thought, but least of all, among the "regulars" in the profession and schools of medicine.

MAN AS A PHYSIOLOGICAL BEING.

Being based upon natural laws, and their relation to the physical organism, Osteopathy, both in its application and results, is largely drawn from physiology. The constituents which compose the material of which the body is constructed are varied in character, and possess substances of which the most expert chemist never dreamed. These agencies are so subtle and as yet so elusive, that neither the microscope nor the finest analytical process have laid them bare, or brought them sufficiently in view, that they may be named and analyzed. No more have all the forces of nature in the elements above and the earth below been discovered, and the proper significance of many of those that are known to exist has not been determined.

It is known, however, that of the understood substances of the human organism, a power is exerted by them, which is simply wonderful—a power which combines the potentialities of the universe, such as growth, motion, assimilation, excretion and reproduction, and these, so to speak, are indigenous to the body. They sprang into being with it; were the underlying causes of its growth, and can never be detached as a part and parcel of its structure.

These substances do not depend upon material aid, such as are supposed to come from drugs, for nature being so much wiser than man, she did not need his help in preparing her plans, and in the execution of her purposes. Her plans were formulated, and the power to carry them into effect was eternally and inseparably imbedded in the plans. Each atom at the time of its inception became instinct with the grand purpose of its creation, and proceeded to the fulfilment of its mission. The functions of life, as is evidenced by the most advanced physiology, depend upon the physico-chemical laws for their exercise and maintenance. Remove water, heat and oxygen, or light, air and moisture, and the functions of life cease. In the absence of these agents, the vital functions become inoperative, and consequently non-generative. The human mechanism thus dismantled and deprived of its life-force, is reduced to not only a state of inertia, but to absolute death. The conditions of physical life are as definitely fixed and controlled by physico-chemical laws as the atoms in the material universe are definitely operated upon by gravitation. The potency of the one, in the sustenance of life, and keeping in motion the atoms which play such a conspicuous part in the machinery as a grand whole, is as great as the other in its tendency to move one body towards another.

The supremacy of the natural laws is above all and over all, and no human invention, or combinations that man can make in the broadest fields of pharmacy, can supplant nature. The apothecary may, in the preparation of a lotion or anodyne, imitate to a certain extent the chemical substances of nature, but having been compounded by him, these are only an imitation, and when applied externally or taken into the system, they are foreign and unnatural substances, and as such, they will neither assimilate nor act in perfect harmony with the forces which are already there and at work. The application of the natural laws of vital force is the mission of Osteopathy. It came into existence for this purpose, and is determined to give this living force free course in the functions of the human organism. It makes no difference how much these natural laws may vary in effect in health or disease, their efficiency cannot be augmented by drugs or the compounds of the chemist. Nature is her own physician, and her prescriptions cannot be duplicated by man. Drugs are

productive of nothing that contain life. Not an atom of
nutriment do they create in the body by which they are
absorbed; hence, being like so much dead matter, they
have no influence over the physiological conditions which
make and perpetuate life. The origin of man may be
traced back by Prof. Huxley and others, to the lowest
bioplastic status, and even then, its germinal stage will not
be found to have been the result of mechanical motion
or chemical combination.

Nature asserts her supremacy throughout all living
organisms; from the smallest and most insignificant to the
greatest—up to man. All were fashioned by the same Almighty hand, and endowed with functions which sustain
life. Each having no more than were necessary to the
fulness of existence, and none any less than were absolutely
required. Chemical synthesis does not and cannot unfold
or produce life. The manufacture of the simplest ferment
by the chemist is just as utterly impossible as it is for him
to create an entire living machine. "If life, even as it exists
in the formless bioplasm which precedes the cell, cannot
be referred to physico-chemical conditions, then Haeckel's
materialistic explanation of the origin of man's body falls
to the ground.

"The theory is still more untenable when applied, not
simply to the production, but to the formation and speciali-
zation of organic life. Physico-chemical conditions may
indeed exert an influence on its cohesion and unity. This
demands a directing thought which shall determine the
development of the living being by harmonizing its various
elements, with a view of the whole.

"The power of synthesis which converts external matter,
and produces organic substances, resides inherently in the
bioplasmic basis of the body. This growing action, obeying
the controlling power, shows itself in a complex machine

like man, and this machine or intricate organism, is an assemblage of cells, in which the life of each element is maintained, and these cells form themselves into groups of tissues, organs and systems.

"Certain animals and vegetables are so much the creatures of atmospheric changes and conditions, they cannot live without them, while man possesses within himself the life-giving, life-sustaining elements."

Man being the highest manifestation of God, both in his physical and intellectual construction, he was more generously endowed than all other created things with inherent powers of reproduction of the elements that sustain life. His reasoning faculties make him less subject to the influences which may affect the animals, and at the same time make him less dependent upon shifting cosmic conditions. The instinct of the animal consigns it to live in certain grooves, or channels, which it naturally but unwittingly follows. It lives upon certain vegetables if it be one of the herbivora family, guided in its choice by the unerring impulse, or instinct, without an apprehension as to the results. Man, however, can adapt himself to either animal or vegetable diet, and whilst he is endowed with a mechanism that is common to his race—a mechanism that is duplicated in every human being with whom he comes in contact in every nook and corner of the globe, he chooses for himself, rejecting any article of food that he pleases, not because it is poisonous but because it is not palatable, or craved by his appetite. While man's general bodily conditions may be more or less affected by external circumstances, the internal organism of his being—that is to say, its vitalizing atomic forces, remain the same, and are constantly working out the problem of life, renewing their own existence, and throwing off dead and effete matter.

External variations, if any, are compensated for by the

equilibrium of the nicely adjusted and perfected organism within. The nervous system acts for the machine of the body like the lever or index to a watch, which controls the effective strength of the hair-spring, and thus regulates the vibrations of the balance. The nerves regulate the harmony between the conditions essential to life. When this concord is destroyed, Osteopathy teaches that the natural law of nerve force has been obstructed, and that when this impediment has been removed the nerve force will become normal. In its contention and successful demonstration that the nerve force will become normal when the impediment is removed, Osteopathy has achieved a distinction as a therapeutic science. When viewed from a physiological standpoint, we find in man that every part of the organism is linked and adjusted with reference to the whole. We also find the division of labor applied by law to the various parts and functions of the machinery, and further see that the machinery as a whole is in accord with the physico-chemical laws which regulate the life of the material universe.

By virtue of its miraculous construction, the human body maintains a self-poise which is indispensable to its independence. As we have already stated in a preceding chapter, every principle of mechanics and physics is conspicuously exemplified in the human machine. The reserve forces of the body are immense, and are manifested in its power of recuperation, when the laws of nature are not obstructed in their operation. The introduction into the human machine of combustibles produces a similar effect to that which moves the steam engine. These combustibles generate heat, a part of which is transmuted or changed into work. Much of this work is absorbed by resistance. "In this respect, the human machine surpasses all other mechanisms produced by industry. In fact the work of this machine can rise to the fifth of the mechanical equivalent of the heat produced, while other machines hardly obtain the half of these results."

"The stomach which acts as the retort of the body, dissolves the materials which are taken in, and these flow into an elongated tube. The force-pump and suction take up the blood, which waters its feeders, its springs, pistons and wheels. The principles of hydraulics are seen in the circulation of the blood.

"The nerves serve as reins and spurs. The nervous system forms the compensating fly wheel of the machinery, balancing losses and gains. * * * Water being an indispensable element in the constitution of the surroundings, in which the living organs are evolved and perform their functions, there ought to be found in the body such a general structural disposition as will provide for the regular maintenance of the necessary quantity of water in the system, whatever losses and gains occur. Accordingly we find just such an arrangement—an apparatus which provides for the loss and restoration of the quantity of water in the system, and it is very complicated, involving a number of processes, such as secretion, exhalation, circulation, etc.; and thus is maintained the presence of water in a certain, definite proportion, in the internal organism, as the condition of the vital functions."

The comparison of the human machine to the locomotive engine is a good one. There are many points of agreement between the two; but in fineness of construction, complication of the machinery, and the adaptation of its parts to each other, the human machine is as infinitely superior to the man-made mechanism as the beauty and fragrance of the rose are superior to the artificial flower. Solomon in all his glory, possessing all the riches that man could desire, by which he could command the combined skill of all the master workmen, artists, and mechanics of his day, could not array himself like one of these natural flowers; that is to say, he could not produce the lily; could produce only an imitation. The man-made machine, though marvelous in its inception, combining in its construction the greatest skill and wisdom that can be or has been attained by the finite mind, is but after all a coarse, crude likeness of nature. Take for instance the eye, one of the most delicately made organs of the human machine. Man has never approximated its internal wonders, or even outward appearance so far as the living eye is concerned. Think of the eyelids and their functions! of how they are formed; think of the eyelashes and the purpose for which they were made! their tactile sensibility! the Meibomian glands, the lachrymal gland, the orbicularis palpebrarum muscle, the muscles themselves, the optic nerve; the eyeball and its object; its movements; the function of the sclerotic coat; the cornea which serves as a window to the light; its corpuscles; its nerves; the choroid coat; the ciliary processes; their muscles; the iris and its functions; the muscular fibers; the pupils and their movement; the blood vessels and the nutrition they furnish; the veins; the oculo-motor; the sympathetic nerves; the retina and its functions; its chemical reaction; the rods and cones; the blind spots; the macula lutea, the visual purple; the aqueous humor; vitrous humor; crystalline lens; the way in which images are thrown upon the retina; the power of refraction; dioptric media; intraocular pressure; etc., etc. And yet, these are but an insignificant part, so far as numbers are concerned, of the different portions of the human eye. Man can never copy it; he can never make the eye as God has made it. He can only look at and be struck with wonder at its marvelous workmanship.

Take a single nerve out of the almost innumerable num-

ber which are found in the system and analyze it so far as
we can. and what is it? It is composed of the primitive
fibril, the naked axial cylinder, the clothed axis cylinder,
covered by the white substance of Schwann; the clothed
axis cylinder covered by the neurilemma, and the clothed
axis cylinder with both these coverings. These are the
different parts of a complete nerve. Going deeper into
the subject of nerves, we find there is a difference between
the fibers of the cerebro-spinal system and those of the
sympathetic system. Then there are the nervi-nervorum;
the tensile strength of nerves; the centripetal and centrifu-
gal divisions of nerve fibers; the white and gray nerves;
the functions of these nerves; their velocity, etc. Take
any part of the machine called man, and its construction,
and it furnishes food for thought and contemplation that
will consume the better part of our natural lives to comprehend, and even then we can get but a glimpse of what it
really is. New discoveries are being constantly made as to
natural laws and their effects, but the field is big enough
and so prolific that man, although he may ever extend his
researches, will have never discovered all there is concerning the human machinery. The general features of man's
"make-up" are known and their constituent parts fairly
well understood, but there are finer mysteries to be explored
and explained, and grander problems to be solved. In the
solution of these problems, the skilled Osteopath will play
a conspicuous part. Although the time has been brief
since he has come upon the stage of action, as the advocate
of a new method, and as a champion of the new science,
he has, by his intelligent application of his knowledge of
anatomy, revolutionized the world of thought in the helpful and healthful results which have followed. He has
shown that anatomy means something, and that if deeply,
closely and persistently studied, and its principles are prac-

8

tically carried out in the healing of diseases, that the demands of nature are fully met and satisfied, without employing drugs as auxiliaries.

The Osteopath has called the attention of the world to
the fact that anatomy is not simply one of the studies included in the course at a medical college which must be
perused, because a slight knowledge of it must be obtained
before a diploma will be given, but he has shown that it is
of prime importance, and constitutes one of the foundation
stones upon which the art of healing is based. Neither
the practitioner of medicine, nor any other man who as-
sumes the role of a doctor whose province it is to administer to disease and heal the sick, can do so intelligently,
and with a proper understanding of what he is doing, and
of what he ought to do, without possessing a thorough and
practical knowledge of anatomy. Along with this study,
the Osteopath takes as an inseparable companion, physiology, which, through his efforts, is becoming more and
more important in the art of healing.

The more intelligent and cultured portion of almost
every community is growing tired of drugs in the practice
of medicine. Homeopathy was a step in advance of the
old school; that is, in the direction of diminishing the
quantity of drugs that were thought to be necessary to the
well-being of a patient. The tendency among all well-
educated physicians of all schools that teach materia medica is to lessen not only the dose, but to give drugs less frequently, and those that are less powerful in their effects.
The M. D.'s are beginning to appreciate the fact that a less
reliance upon drugs, and a greater reliance upon nature
is the most sensible method of procedure, and this idea is
obtaining more and more among all classes of physicians.

Scholarly and scientific attainment will compel the

medical schools to reorganize their course of study, so that the disciples of the healing art may be in more thorough accord with the philosophy as well as the practice of the laws of life. A perfect knowledge of these will place the practitioner upon higher ground, and give him an insight into the needs and requirements of suffering humanity that he never had before. It will gradually dawn upon his mind that nature is the greatest physician, and that when her recuperative forces are permitted to have full sway that disease can be more successfully battled with and baffled than it can be by a resort to the preparation of the chemist.

The potent factor which is leading the van in the onward march of events, and the one that is destined to change public and professional sentiment regarding the old methods of the treatment of disease, is Osteopathy. It has entered the arena of combat, as a young, stalwart knight, equipped with the sword and shield of truth, and, like Ivanhoe, returned from the Holy Land, is unhorsing man after man among the proud templars who dare to oppose its onward sweep. This new science, like many others which have been discovered will, like them, require years and years of close thought and practice before its great value to man will become fully recognized. A new truth is generally slow of development. The antagonizing forces which are naturally arrayed against it have the prestige of age behind them and are strong in numbers. Its converts are as yet few and scattered; these are however, aggressive in thought and spirit; they are aggressive, because they love the brotherhood of man. They are animated with a desire to help their fellows, and to help them in the easiest and simplest way that has been suggested by nature. That Osteopathy will eventually obtain a per-

manent foothold among all classes of people, we have no
doubt. Its apostles are increasing in both numbers and
zeal. The more they learn of Osteopathy, and the more
they see of its effects upon life and health, the more deeply
convinced are they that it ushers in a new era in therapeutics.

Wherever the skilled, conscientious Osteopath flings his
banner to the breeze, it will not be long before he will receive a growing recognition from the community which
surrounds him. There are cultured men and women in
all localities, it makes no difference what the trend of
thought has been on medical practice or professional teaching, who think for themselves; men and women who have
become surfeited with even the smell of drugs when used
as remedies for the sick and the afflicted, and will gladly
aid in forming a nucleous to the new thought. Each student armed with a diploma which he has won, because he
merits it, and which has been given him after two years
of hard study, goes forth as the herald of a better way to
life and health. The success which will follow his efforts,
when properly applied, as certainly as day follows night,
is bound to arrest thought and attention among the
thoughtful; among those who are not so wedded to their
predilections that they are unwilling to at least recognize
the existence of plain, honest facts when they see them.

The following story, though told in poetic strains, is one
among the many hundreds of cases that has been successfully treated at the American School of Osteopathy, which
is located at Kirksville, Missouri; it is entitled, "A CASE
OF NINETY-THREE," and was written by N. J. S.

This is no fictive sketch to touch the reader's mind with
    doubt,
A thousand persons know the facts I herein rhyme about.

Her name was Brown, she was a wife, her home was down
    in Pike,
The land "Joe Bowers" departed from to make a minstrel
    strike.

She came a helpless invalid to Kirksville in the fall,
St. Louis and Chicago drugs had helped her none at all;

That dread disease, paralysis, her limbs had fettered fast,
And she had lacked the power to walk for many seasons
    past.

She was not old, life's springtime touched her most res-
    ponsive heart,
And nature lent her sweetness far beynod the show of art;

Hope set affliction to her view in optimistic light,
Her words were never petulant, her smiles were always
    bright;

Her eyes were blue as summer skies, her voice was soft
    and low,
Her hair caught sunlight bright as that where rare magno-
    lias grow.

A paradox was hidden in her gentle soul and form—
The weakest of our company, she took our hearts by storm.

She had known wealth and luxury, and happy was, I ween,
Until disease assailed her life at soulful seventeen.

She had an elder sister whom paralysis struck down,
Whose case was long called hopeless by physicians of re-
    nown,

Although they issued out their drugs, as if to try their
    worth,
Until, like summer's tranquil eve, she faded from the earth.

The younger sister far and near consulted big M. D.'s,
When in the icy clutch of that implacable disease;

On cause and treatment they of course failed wholly to
   agree,
But all alike gave out the drugs and charged a royal fee.

Each when he failed would shake his head, with much of
   pompous grace,
And prove it was heredity by her poor sister's case.

An idolizing husband had put climates to the test,
A loving father's money had been used for east and west;

In spite of airs of health resorts and doctors' drugs and
   talk,
She grew no better, grew no worse—she simply could not
   walk.

She bravely hoped the best, and yet as bravely lived re-
   signed,
An invalid in body, but almost a saint in mind.

The fame of Dr. Still reached Pike, in passing on its way
To fill the "regulars" throughout the world with sore dis-
   may;

Then failures of the past rose up to preach the doctrine of
   despair,
And indecision kept her long within her weary chair.

Brown came to Kirksville to inspect the work then turning out,
And left a convert in two days without a shade of doubt.

He, with his wife, full soon returned, remaining quite a
   week
To see if Osteopathy would blanch her tender cheek,

And as he went away he heard such words of hopeful joy,
He quite forgot his thirty years and wept just like a boy.

The days passed on, we marked the change with interest
    intense,
And knew improvement by the cheer her manner would
    dispense.

Weeks multiplied to months, but these, indeed, were all
    too few
To know the rare companionship, we in the patient knew,

When one bright morning word was passed, with most tri-
    umphant air,
"Say, don't you know that Mrs. Brown unaided left her
    chair?"

She sat at breakfast, in her eyes the mist of happy tears,
And gayly cried, "'Tis very true—the first time in long
    years!"

No more she was the invalid, imprisoned, still and weak,
Through all her form the glad life leaped that blossomed
    on her cheek;

She went and came, she laughed and sang, she lived the
    soul of glee,
As blithe as any captive bird that ever soared out free.

She left us soon, returning to her lovely home in Pike,
And sadly we agreed we had small chance to meet her like,

She lightly walked an honest mile to reach the Quincy
    train,
As gracefully as summer flowers wave o'er a breezy plain;

She smiled a pax vobiscum as the train departed east,
And—well, her faith in [1]Still, I'm told, to this day has
    increased.

[1](Note) Dr. Still.

ANATOMY AND PHYSIOLOGY.
THEIR IMPORTANCE IN OSTEOPATHY.

Galen, the ancient Roman physician and anatomist, said: "Every bone of the human skeleton is a verse, and every joint a stanza in a hymn of praise to God."

It is with something akin at least to the spirit and intense love that prompted the above sentiment, that every student and practitioner in the science of Osteopathy should regard anatomy, especially that far more important branch of it which treats of the form and structure of the different parts of the human body. In its widest sense the word skeleton means the whole connective tissue framework, with the integument and its appendages. The word itself, however, when preceded by the term human —as human skeleton, is so suggestive that anatomy to many persons means something exceedingly dry, or ghostly, and for this reason the word to most minds signifies a slightly repulsive idea. When we unexpectedly see the remains of a horse, or the remains of any other animal, which, when living, had been utilized by man in some way for his use and purposes, the sight produces more or less of a shock. When we suddenly view the remains of a human being, that is to say, the skeleton, the sensations we experience are still more intense, for they evoke not only a weird feeling, but a keener and higher conception of what we see. Let all students in anatomy (and all Osteopaths are students) dissociate from the word skeleton its uncanny import, and cultivate a similar fondness for the study that animated and inspired Galen.

He studied anatomy, not simply because it treated of the structure of the matchless framework of man, but

also because the science brought him in thought nearer to the great Architect who planned, builded and set in motion the wonderful mechanism which elicited his profoundest admiration. He saw in every bone a verse, and in "every joint a stanza in a hymn of praise." He dignified anatomy with the supreme importance it merits, notwithstanding the fact that it was in his day a crude science. Could he have lived in the present age, and witnessed the advancement that has been made in the study, his intense regard for it would have been heightened by many degrees, and his hymns of praise would have been sung in more exalted strains. Anatomy and physiology go hand in hand, and one cannot be correctly and practically interpreted without the other. The former tells of the form and structure of the body, and the latter of the uses of the parts and the ways in which they work. In other words, physiology may be regarded in the light of a chemical laboratory, in which, exactly the same changes are carried on (only in an infinitely more perfect manner) than are produced by the working chemist in his shop. One furnishes the nomenclature of the bones, their conformation, location, and in fact everything concerning the general "make up" of the body, while the other discusses the uses of all the internal appliances which work the machinery. Physiology is the extension, expansion and culmination of anatomy, so that a perfect idea of the human organisms cannot be obtained without a thorough knowledge of both.

Lying then, as they do, at the base of all information upon which the science of Osteopathy is predicated, they must be comprehended in all their details. The exterior of the human frame may be seen at a glance. Its shape, its angles, its roundness, its symmetry, its proportions, its irregularities, and in fact, the tout ensemble of the machine as it appears externally. Any one possessing two

good eyes may take in these general features. No knowledge of anatomy is required to do this. But as to the per-
fectness of these features, their normal or abnormal pres-
entation, is another question, and this could only be
decided by a knowledge of comparative anatomy. The
beauty and harmony of the visible side of man can only be
determined by the well-trained eye of the artist, who necessarily needs to combine with his profession the study of
anatomy. It should not be so comprehensive as the knowledge of the surgeon, or as that acquired by the Osteopath,
but it should be extensive enough to enable him to correctly interpret the forms in the models, and to accurately
portray them. The sculptor, perhaps, in the legitimate
pursuit of his art, requires a more thorough understanding of some portions of anatomy than any non-professional
man does. The bones, muscles, arteries and veins are especially studied by him. The artist—the sculptor, should
take a course in the dissecting room if he wishes to perfect
himself in his profession. He can never learn the shape
and articulation of the bones, the origin and insertion of
the muscles, and their relation to each other, without he
does this. A living model is never so comely or so complete
in its external conformation and outlines that the eye of
the finished sculptor will not detect some fault—some inharmony that is quite apparent to his keen observation.
He could not do this, nor neither would he know how to
remedy the defect, without a knowledge of anatomy. The
Greek artists, although they handled the chisel with con-
summate skill, never obtained their ideas of the human
form in the dissecting room. They had before them the
living nude, from which they drew their inspiration. But
the Greeks were of rare beauty and grace. Climatic influences, customs and habits, athletic games and other
amusements were indulged, not only because they were

fond of them, but because these things brought their physiques to the highest degree of development. The Greeks deified themselves. They never looked to God for their ideals, but to man alone. They saw God only through man. He formed their highest concept of God. There has been no race of men and women who were as perfect in form and features as the Greeks. Had the Greeks, however, studied anatomy from the cadaver as our expert artists do to-day, the almost faultless proportions (as we see them who are not artists) that have been transmitted to us in marble, would have been still more symmetrical and complete.

So should the painter who represents living objects, whether they be man or beasts, upon canvas or plaster, be familiar with some of the branches of anatomy. The embalmer would find the same study of immense value to him, as would also the taxidermist. If our knowledge of anatomy can serve us in so many ways as an auxiliary to the more perfect understanding of the simple arts, which are followed as livelihoods by so many persons, how infinite in importance should it be considered by the student, who intends to adopt the practice of Osteopathy as is his life-work.

A mere smattering or superficial knowledge will not do. The gardener who understands his business will not be satisfied to simply cut the weeds down. He persists in his labor until he gets at the roots. When he does this he becomes master of the situation. When the student descends to and successfully grapples with the roots of the study, whether of anatomy or physiology, he is then prepared to build successfully from the foundation up. If he does not do this, he fills the role of a quack and charlatan, and imposes himself upon the credulity of a confiding public without sufficient warrant or justification. Besides

this, he brings the science of Osteopathy, and those who
legitimately practice it, into bad odor. Herein lies one of
the great dangers, or more properly speaking, stumbling
blocks, in the way of success to any newly discovered theory, whether in art, science or invention. Whatever value
they possess is quickly seized upon, appropriated and utilized, by the designing and unscrupulous, to further their
own nefarious schemes. Hence the imitators and constantly increasing race of quacks, who boast of skill not
possessed. Hence the vampires, who, like leeches, fasten
themselves upon every enterprising community, and bleed
their dupes not for the glory and advancement of science,
but for the money that they may obtain. Their only object is to "feather their own nests," and the ultimate result
is that they "befoul their own nests."

A profession and its merits are largely judged by the
common people by those who represent these professions.
How necessary it should be then, that those who are engaged as the practical exponents of these professions,
should be so skilled in their work, and otherwise so careful in their conduct, that the patient, clientele, or public,
may have no just cause of complaint; above all, never knowingly do anything that would fasten upon yourself the word
quack, which is an opprobrium, that every genuine, honest,
professional man will despise.

The wrecks of many a just and praiseworthy cause have
been strewn along the pathway of time, simply because
many of their sponsors and advocates were mere pretenders.
Men who had donned the togas of those who had projected
the thought, like the fabled jackdaw, strutted for a
brief season in borrowed or pilfered plumage. The majority of mankind, in their ignorance, are more ready, perhaps, to join in the condemnation of that which has real
merit than they are to uphold and sustain the same, pro-

vided it bears upon its face the semblance of quackery, which has been placed there by the unworthy means which pretenders employ. The hue and cry of an indiscriminating public have hastened the untimely end of many a good cause, for the reason that those who were its professed followers acted in such a manner that they brought reproach upon both themselves and the cause they represented.

We see this idea existing in the minds of numbers of persons in reference to Christianity, but especially in reference to the churches or denominations which represent Christianity. If a minister of the gospel acts in such a way as to bring obloquy upon his own name, those who are inclined to dislike anything of a religious character, and those who cannot make the distinction between individual responsibility and the cause represented, will make the act a pretext to condemn Christianity. The underlying principles of the church universal cannot be affected by the misconduct of one or a thousand of its ministers, but at the same time their reprehensible conduct may and does to a certain extent retard the growth of religious feeling in the minds of the classes of which I have spoken. One unworthy Osteopath, one imposter, may do more to impede the advancement of the science now than a score could do in ten years hence. It is while the study is in its infancy, while it is struggling for that recognition which all new sciences must fight for, that its friends should redouble their efforts to promote it. When it has shown what it is; when it has attained a permanent position in the confidence of the people, as a profession which may be relied upon as embodying all that is now claimed for it, then the influence of the mountebanks will be less potent than it is to-day.

The importance of being well-grounded then, in the

bed-rock principles of Osteopathy, is apparent to every one
who intends to practice it. The more thorough we become the more respect will be paid to our profession. The
higher we advance the banner under which we have con-
scientiously enlisted, the more numerous will be our
friends. The banner must necessarily gather about it those
who are able, true and strong, and they will help to lift it
still higher and higher, until all who are afflicted may be
able to see its beneficent folds. Anatomy and physiology
then, with the concomitant sciences which form the basis
of osteopathy, should be absorbed by the student in no
half-hearted way. He should probe their mysteries with
all the eagerness and anxiety with which the miner delves
after the precious metals, and with a fixed, positive determination to completely master them. If he does not do
this he is attempting to build upon sand, and no superstructure reared thereon can be either lasting or substantial.

He must bear in mind the truth of what Pope has said:

"Learning by study must be won,
It ne'er was entailed from sire to son."

He cannot pick it up in a promiscuous manner. He
cannot inherit it. He cannot hire another to study for
him. Daily intercourse with a skilled Osteopath and having carte blanche to his operating rooms, whilst patients
are being treated, may furnish him with some practical
knowledge. But all this will not supply him with what he
needs and must have in order to become a competent and
perfectly reliable Osteopath. He must, like the gardener
of whom we have spoken, get at the root. Learn the theory; understand the whys and the wherefores, and then
put to a practical test what he has acquired in theory.

There is no way by which he can dodge what has been laid down in the text-books. If he does attempt to avoid it, he is an unwilling student and will always be an unskilled and an indifferent practitioner.

It is the man or woman who is in love with a profession that becomes proficient in it. When love is the incentive it imparts an all conquering power that will surmount all barriers that lead to the object to be attained. "Love," we are told, when considered in the light of a tender passion, "laughs at locksmiths," and will find a way out of the most trying difficulties. The love of Osteopathy or any other science that we have taken up will not only push us through, but will cause us to study it and cogitate over it until we have mastered it.

We have in mind an illustration. We knew a little girl of about twelve years of age. Her mother being anxious that she should take music lessons and learn how to play on the piano, employed a teacher. The teacher came to the house and patiently gave her lessons for six months. The girl expressed her repugnance at the arrangement that had been made by her mother, but took the lessons. Her heart, however, was not in her work. Her lessons occupied a half hour each, but she was always impatient of the time, and would often look at the clock and say: "Such a long half-hour! Will I never get through?" The result was that she learned imperfectly some of the rudimentary principles of music, but after a brief time abandoned all and gave it up entirely. She had a brother two years younger than herself. He expressed a desire to take lessons in music on the piano, and the same teacher was employed. He never could practice enough, and was at the instrument every opportunity that was offered. Need I say what the result was in his case? He became one of the most expert performers on the piano in that entire region of country. He loved that in which he was engaged. That which had been a burden to his sister was a pleasure to him. How often do we see it the case of persons having studied a profession and practiced it a few years, and then give it up for something else. What was generally the cause? "Why I never liked it! Father wanted me to study that profession and I did so to please him."

If you wish to become a good professional man, fall in love with the profession that you adopt, and you will move along like a ship under a pleasant breeze. Despise what you have undertaken to do, and generally speaking, you will not be compensated for the time that has been wasted in its accomplishment.

After the principles of anatomy and physiology have been mastered, there will follow chemistry in its various branches, symptomatology and physical diagnosis, and all other studies that are included in the curriculum of a first-class Osteopathic school.

It should not be thought, however, by the student, after he has received his diploma, that there is nothing more for him to do but to throw his shingle to the breeze and begin and continue the practice. His diploma does no more than open the door through which he passes into the great field which is to be the scene of his future investigations, triumphs or failures. He is no longer under the immediate eye of the professor, nor is he confined to the narrow limits of scholastic walls. His campus is the world, where his acts will be closely scrutinized by a curious public. Every act and everything that he does, whether as a citizen or professional man, will be placed to his credit or to his dishonor. He has laid the foundation, and begins to rear the superstructure. The public will witness the manner and the progress that is being made, and watch the building with increasing interest as it goes up. Self-ap-

pointed committees will inspect the work, and offer from
time to time such advice as they think the necessities of
the case demand. He will have much to contend with,
and will often find that while he thinks he has completely
mastered the theory that there is still a wide difference between theory and actual practice. He receives additional
light from all sides. Each new case will furnish new
thoughts. Each new patient will present something dis-
similar to the preceding one, and while he may be conver-
sant with the general features of all ordinary diseases, there
may be complications which are unusual. He will have an
opportunity to study man as he never had before, and the
different types of men, their mentality and temperaments,
and these varied attributes will enter more or less into
what he does, how he does it, and to what extent his diagnosis or investigations will lead him. A disease may develop the same characteristic traits in man generally, but
the method of treatment, if the practitioner fully understands the nature and disposition of the patient, will not
always be the same—will be more or less heroic, or more
or less gentle, continued or abated.

Then there will be presented cases which will be entirely
new both in theory and practice; something that the physician knows nothing about. The symptoms may indicate
half a dozen diseases, and yet not conclusively point spe-
cifically to any one. Under such circumstances the practitioner will, of course, be left entirely to his own judgment. At such a crises he will realize the importance of
what he has stored away in thought. He should always be
on the alert, be careful of what he gathers, so that it may
fructify into something that he may use for the good of his
profession.

The Osteopath, however wise he may be, however much
he may accumulate in the way of theoretical and prac-

9

tical knowledge, his nuggets in this respect may be numerous and costly, yet he will realize that there is much more for him to learn.   Sir Isaac Newton, the great philosopher, in speaking of his achievements, said that he had only been gathering pebbles along the shore of the great ocean, which still lay unexplored before him.   Every Osteopath who is a sincere and honest seeker after truth may pick up a good many beautiful, polished and valuable pebbles along the shores of thought, which trend in his direction, and he may appropriate these to the promotion of the best interests of the noble science which he has espoused.

Much depends upon the point from which the beginning is made.   If you start from the right place the obstacles will not be so hard to overcome.   If you start wrong you will not only squander time, but are liable to become discouraged.   Begin then at the bottom, where the chief, strong work is to be done.   Mix and dovetail in the big and little stones with the mortar, so that the masonry, as a whole, may be compact and so cemented that the super-added weight will only increase the solidity of the structure.

Anatomy and physiology are the nethermost rocks in the Osteopathic foundation, and must be used as the initial stepping stones.   The student cannot mount to a place at the top by attempting to ascend in any other way. If he makes the effort he will have his pains for his folly. Failures in life are more often the outcome of a bad start than they are of accidents or misfortunes.   A man may recover from an accident or misfortune, but rarely ever will he right himself if he takes up the journey of life by moving off in the wrong direction.   There are no "short cut" roads to Osteopathy; no crossing lots to reach it, and the student who tries a different route to that which has been surveyed and blazed out, will find himself in a labyrinth of

intricacies that he cannot successfully travel. The landmarks of the science are not old. They have been but recently established, but so carefully and definitely have they
been located that they mark the only recognized highway
that the seeker after osteopathic information can pursue.

In an issue of a Chicago medical publication the following admissions were made in reference to the importance of
anatomy:

"Briefly, the system of Osteopathy appears to be this,
that the student is thoroughly trained in anatomy upon the
living body, going over and locating the bones with all
their prominences and depressions, then the ligaments
and muscles attached, and the vessels, nerves and other
structures as related to the bony framework of the body.
By this method of training the student is so familiar with
the living human body that he is enabled to detect many
deviations from the normal standard that would escape the
ordinary physician, and which are yet capable of account-
ing for many of the ills that affect the body.

"Now as to the value of this method of teaching anatomy
there can be no question, or of its vast superiority over the
methods in vogue at the medical schools of the present. I
well remember hearing Pancoast urge upon his students
the importance of studying 'living anatomy.' It should
be introduced into the schools of medicine at once, and a
large part of the additional time secured by lengthening
the course should be thus utilized. Frankly, I know very
few doctors from whom a crooked spine or a sprained
joint would receive as intelligent treatment as from an
ideally proficient Osteopath."

MODERN CIVILIZATION.

Each rank and condition in life, from the humblest up to the highest, have their responsibilities, their good and bad features. The highest culture makes its mistakes in certain directions, as well as the densest illiteracy. One tends towards one extreme, and the other as far away in the opposite direction, oftentimes impelled thereto by the force of circumstances. Climate and temperament have much to do with a man's taste and habits, and these in turn have their influence upon health.

One of the most productive sources of ill health to the cultured is their extreme indulgence in the comforts and luxuries of life, without the use of any special labor to attain them. Opulence superinduces this condition. Money will purchase these things; hence money is substituted for the effort or physical exertion which may be necessary to enjoy them. These desires are gratified through the appliances of the arts and sciences which supply without delay the rapidly accumulating demands of indolence and capricious taste. Such methods and ways of living result in the increasing and multiplying of the diseases to which the nervous system may be subjected.

The gently nursed are prone to take the most luxurious care of the body; it is not permitted to come in contact with the life-giving elements of nature, such as the sun, air, ozone and natural electrical currents. The consequence is, that physical deterioration ensues, the vital forces are weakened and the strength of the body is far below what it should be, even in the most ordinary conditions. The physical system has about the same growth in kind that the hot-house plant does, when protected by glass and shade from the sunlight. The flower will bud and bloom in the conservatory and will send forth its fragrance, but its perfume soon dies, and the flower itself will wither and perish, if exposed to the ⬤ or to unusual conditions. Its beauty will be more delicate, but not so vigorous, not so deep and glowing in tint and color. It will, however, be a beauty that will soon fade.

So it is with humanity which lives and thrives only through the hot-house process, and this is the manner in which modern civilization protects and rears its offspring. The specimens may be comely and attractive to look upon, when exteriorly considered; but they are lacking in the sap and vitality and durability of those that have been grown in the freer domain of nature. The latter prove to be the most effective thinkers as well as workers, and do more as factors to promote the furtherance of any cause, trade or profession in which they embark, for the reason that vigorous health gives them a greater incentive and a broader field in which to labor.

The hot-house plan which is followed by the highest civilization leads with almost unerring certainty to nervous affections, which are as varied in character as it is possible to conceive of. These contribute more or less to the weakening of the body as well as the mind, which cannot perform their natural functions under such circumstances. The methods which are employed by those who are gently nursed, and whose means are in no sense limited, to secure peace and rest, are having just the contrary result. Instead of the peace and quiet for which they strive and most desire they are troubled with incessant worry and many ills. Why is this so? Because there is no adequate machinery for a brain and nervous system thus developed, upon which it may act and react. "The stream is too powerful for the banks to hold it; the stream exhausts the water supply and becomes red hot in the apparatus that should give it vent. The muscular, vascular and bony systems are no longer capable of sustaining the unusual strain. Nature is crying out in distress at the unequal burdens thrown upon its sustaining organs of vitality, which are wasted and destroyed by want of proper use, or burned and desolated upon a desert of artificial culture and unnatural heat, the result of ignorance and thoughtlessness of the true ends of life."

The human system, when cared for in the manner indicated, cannot be strong, for it does not possess the necessary braces and girders to support it. Its weakness is manifested at every exposure and the ordinary shocks of life, such as come in the ordinary daily avocations, are extremely annoying and these harrassings result finally in disease. Unstable and decayed timbers are loosely jointed by the careless and inefficient interference of a meddlesome art, which knows no better than to sacrifice everything to whim for the sake of form, or its conceptions of transient and ephemeral beauty.

We do not pretend to say that nervous derangements belong only to those who are raised in a hot-house fashion. All persons are more or less subjected to them, and they may occur in a thousand different ways. What we do say is this: That persons who are so tenderly nursed are more liable to the exhibitions of nervous phenomena than any other class of people. The reason is perfectly apparent, and may be explained in two words—physical weakness—or a too delicately organized system. It is not essential to the strength of the body that it should be unreasonably exposed; there is a happy medium in the things which tend to promote life and health as there is in all other extremes.

Creation is the result of life, as well as are all things

else, and the status of strength and conditions of endurance attained by life come from the law of the survival of
the fittest in the struggles of mankind. The attempt to
aid and foster life by artificial means, without understanding its wants and real necessities, will weaken and destroy
its properties of endurance, and when these are destroyed
we cannot get out of life but a moiety of that which is
worth the living. The heights and depths which are
reached by life are far beyond the grasp of human
thoughts; its duties and activities are continually evolved
and involved in a natural process, and if these are interfered with by ignorant plans the tendency is to handicap or
paralyze the recuperative powers of nature, not being sufficiently supplemented with what should be the normal
strength of the body, they fail very materially of their purpose, and when these already weakened powers are further
shorn of their vigor by artificial methods which are born
of ignorance, nature is to that extent much further ham-
pered.

Life does not work at random, nor does it work automatically, but for special ends. This is seen in both the
brute and plant creation, and in everything into which it
enters. It is ruled by a law of necessity, which is as fixed
in principle as mathematics, and this law pursues its way
of combining and distributing for the sole purpose of the
ultimate perfection of the natural order.

A more artificial condition is met with in man. He is
endowed with the special powers of will, guided by reason,
and is more responsible for the results of his own development. Even in him nature always discovers the weakness
of his plans and masters the situation. Man may, by special training and the full exercise of his will power, change
natural tendencies and control the circumstances of his
own life. He may not only do this to a very large extent,

and thus become really the architect of his own fortune, but he may control the natural and instinctive life of the inferior beings which surround him, and which minister to his wants and happiness.

He is then given the power to regulate, multiply and improve, or the power to destroy all the creatures which live within his reach. In other words, he stands at the head of creation, and has complete and absolute dominion over all other beings lower in the scale than himself. But man is inconsistent. The natural supposition is, that he would master the laws of his own being so that he would thoroughly understand how to live in order to accomplish the purposes for which life has been given him. But he has done nothing of the kind. He has studied with greater care the laws and conditions of the brute and plant creation. These are in many respects less complicated than the laws of his own being, but they are of infinitely less importance to him. Being in the main ignorant of himself, he does that, or neglects to do that, which destroys health and prevents him from enjoying life. The avenues to health are open, and by walking therein he may find peace and happiness. Man alone, of all created beings, has the opportunity to grasp the key which will open to him an earthly paradise.

Being endowed with reason, he can find the way if he so desires, and after he has discovered the way, he can, by the force of his reason, pursue the path which leads him to this goal. His intellect proclaims him to be superior to all other creations, and yet, of all other creations, he is the most unreasonable in following that which pertains to his health and happiness.

The animal with only his instinct fulfills his mission. The man with his reasoning power—the power that can discriminate between right and wrong—between that which is baneful and uplifting, in the vast majority of cases, seldom, if ever, does the best he can; upon the contrary he often does the very opposite to that which his reason dictates. Hence he is consciously responsible for not arriving at the goal which is in his power to reach. He uses his knowledge of inferior life around him for the mere gratification of his whims or appetite, without considering that the laws which govern them may be applied to himself with benefit.

Man could have better understood the processes of his own nature if he had properly heeded the efforts that were necessary to effect the changes in the unconscious world around him. By doing this he would have been more impressed with the importance of understanding the processes of his own generation, and appreciated in a more exalted degree his endowments and privileges. The proper development of the human race and the methods of rational generation which lead to health and happiness are totally disregarded, and the result is, that man finds himself precipitated in a reckless manner into the arena of existence, unmindful of the consequences and without a proper understanding of its physical laws, simply from the fact that he has not studied them independently of the artificial interferences which environment and habits have thrown in his way as stumbling blocks.

Those whose profession in life entitled them to speak with, perhaps, more authority than all others never urged upon the world with sufficient emphasis the prime importance of the matters we have been discussing. These things have been presented by the humanitarians and philanthropists with much zeal along a line of thought that has resulted to a certain extent beneficially; but the good might have been ten-fold greater, if the physicians had as a unit

spoken at all times regarding their supreme importance—
the laws of life.

What are we here for if not to enjoy life, and to do all
the good we can to others? The physician can exert the
most salutary influence of all other men respecting the laws
of health. It is in the line of his business exclusively,
and when he fails to instruct those with whom he comes
in daily contact he is guilty of a dereliction of duty for
which there is no reasonable excuse.

The adolescent period in the life of every human being
is the time of all others, during their earthly pilgrimage,
when they are particularly susceptible to impressions upon
their nervous centers. It is the time when nature begins
and carries into operation her plans and processes of pro-
creation. It is a period of extreme danger, and of untold
importance to health. It is a period when strength may
depart; when life thereafter may become a burden; when
the mental as well as the physical organism may suffer and
will enventually render them entirely useless as either
thinkers or workers in the great busy world.

It is a time when young manhood and young woman-
hood may get a set-back from which they may never recover. If the imagination which controls many reflexes of
the great sympathetic system is not curbed and set in its
proper channel the most disastrous consequences, as we
have already stated, may follow. These unfortunate results are more apt to follow the arrival of this period in
the hot-house plants of humanity, who are already enfeebled by undertoned and undeveloped powers of life than
in others. This is natural and this is nature.

The weakened powers are called upon by the sharpened
appetites of a nervous system out of all proportion; this
excessive growth of appetite is the outcome of the modern

school of forcing, common as habits of society, and the wreckage and disaster is fearful to contemplate.

Of all the periods in life then that known as the adolescent is the most important, and of all the periods in life that is the time when those who are passing through it should receive instruction and particular warning about their true natures.

If the laws of our physical and moral being were properly understood the marriage relations would be fruitful of a complete transformation in the natures of mankind. There would be less disease, a fewer number of deaths resulting from disease and far greater happiness; because then man would have the opportunity of freeing himself from the ills of heredity, which he is now doomed to inherit from misguided parents and ancestors.

A new race would appear with contented, loving minds and dispositions, and work out more easily the problem of life.

The subject of heredity and its influence on disease is of so much importance that the osteopath should study it deeply. Its tendencies are hard to control, especially so if the patient has taken the traits without any attempt to resist their influence. We mean those hereditary characteristics that tend to evil, and if indulged, will produce ill health and unhappiness. "It sometimes happens that the inheritance of physical and mental weakness is so pronounced that health and strength are as much unnatural states as disease itself. Though the osteopath may succeed in removing all obstructions to the natural flow of blood and the fluids, there is a birthright of imperfect organization to contend against which cannot be overcome. This is one of the causes why many diseases have been pronounced incurable by the medical profession." The Osteopath, however, in nearly all cases which effect the nerves,

may touch the "key-board of physical harmony" and make
the currents of life flow in their proper channels, and in
something like their normal volume. Nature is always
wide awake. Her laws may be so often and so wantonly
broken that her recuperative powers become greatly weakened; even then with the aid of an expert Osteopath, she
may perform wonders. While to past and present civilization are due the innumerable blessings which are enjoyed by mankind to-day; while it has taught us all that
we know of good; given us a higher conception of life, a
broader view of our relations to our fellow men, and is forever evolving some new truth, applying some new principle in the arts and sciences for our benefit, that we may
be wiser and happier—while it has done and is doing all
this it has many mistakes to answer for; mistakes which
are traced only to its door, the chief of which is the one
with which we have been dealing—the mistake of raising
its offspring as it would a hot-house plant.

I speak more particularly with reference to those who
are supplied with gigantic fortunes. Money is all right,
and money-getting is all right, when it is honestly obtained. But the luxuries which money purchases, which
are had for the asking, are not the blessing they would be
if they came by a little labor and exertion. Great wealth
and high attainments naturally suggest a high state of living; luxurious habits are easily adopted because they cost
no labor or exertion. Great wealth suggests a life of ease
and indolence which can be pursued at the taste of the
one so inclined without any self-sacrifice in the way of bodily activity.

It is so much more convenient and so much easier to
have done or hire done that which requires labor either
physically or mentally than to be subjected to the doing
of it yourself, and since you have ample means at com-

mand—means that can employ the most skilled laborer, why not have it done?

Inertia and luxurious habits of living entail upon the offspring weakened constitutions, and the greater the physical weakness the greater the tendency to nervous diseases. These habits which diminish the physical powers militate against the full development of the body in stature and roundness and firmness of proportion. Take fifty thousand people who indulge luxurious habits and expend no physical labor towards attaining the objects they have and enjoy. Take another fifty thousand people near by who enjoy life and many of its luxuries, but to get these they have undergone a reasonable amount of physical labor and exercise. Which of the offspring of the two classes mentioned will be the more healthy, taking it for granted that the parents of each are equally vigorous? Which will be, generally speaking, the larger race of the two? Which class will embrace the more manhood? Which will achieve more in life? Which is destined to live the longer? Which will be the more subject to disease and succumb quicker to its inroads? Can there be any question?

But continue this for three or four generations, and what will the harvest be? One will dwindle to a race of pigmies both in mind and body, and the other will grow to a race of giants, comparatively speaking, in both mind and stature.

There is no science that so fully teaches the value of these lessons as Osteopathy. Based upon anatomy, it is bound to look deeper than the mere surface of things, and draw its conclusions as to the future results of what may be seen and understood from a profound knowledge of what it grasps. Being founded then upon anatomy that study is made the chief stone in the structure: Each student becomes so familiar with its principles by hard study both in and out of the dissecting room that he can at a glance detect any abnormalty by the mere touch of his fingers, and without the aid of his eyes, provided this abnormality is developed upon the surface of the body. He knows the construction and conformation of the human frame so thoroughly that any unnatural growth or development is not only quickly discovered, but he understands the causes which contribute to the malformation. He also understands whether or not the different parts of the body are developed as they ought to be, whether there is an undue shrinkage of some of the organs and an abnormal growth of others. He knows so intimately the relations that one part of the human system bears towards another that he has but little, if any, difficulty in making his diagnosis, and then proceeding in an intelligent manner to right the wrong and readjust the machinery. So familiar is he with the anatomical construction of the body that he can tell in the dark with his skilled fingers the difference between a well-developed form and a poorly developed form. The bare touch of the surface is all that is required to convey to him what is lacking or what has been over-developed. In other words, the human frame, with its bones, its hard and soft tissues, is the key-board which responds to his touch with as much certainty and correctness as the key-board does to the touch of the expert musician. Its inharmony is as apparent to him as it is to the musician.

OSTEOPATHY IN HARMONY WITH THE BIBLE.

Osteopathy is not allied in any sense of the word with spiritualism in any of its teachings, nor has it any bearing upon or connection with occultism as practiced and taught in the Middle Ages. Nor has it any power derived from what is claimed to be supernatural agency. Neither does it deal in necromancy, hypnotism or witchcraft. But its methods are fixed and determined and wholly built upon plain, common sense duductions, founded upon the laws of nature, aided and supplemented by a thorough knowledge of anatomy and physiology. These two studies furnish the key which opens the way to health and life, and every skilled osteopath who is well grounded in these two sciences can achieve results which follow only from natural causes when his knowledge of these studies is practically and intelligently applied.

There is nothing hidden or mysterious in the practice. It is as open as the light of day and as simple and as easily understood as are the pages of a familiar book. The Osteopath is not an inspired magician from the orient, nor is he a descendant of the ancient Persians. Nor do his fingers constitute the spear of Ithuriel, at whose touch the human body can be instantaneously transformed into shape. He is simply a man possessing the ordinary attributes with which all men are endowed by the Creator, and any man or woman having the usual ability to understand and apply what has been acquired can become a successful osteopath, with this difference, that some persons have what we call more of a natural gift or talent for the business or profession they follow than others. These, as a general rule, become more expert and their efforts are attended with perhaps speedier if not more successful results.

In this connection we will state again (having already done so in the opening pages of this book) for the benefit of our readers, that Osteopathy is neither the massage treatment nor the Swedish movement. The Osteopath is compelled to obtain a thorough knowledge of anatomy and physiology with their adjuncts by an attendance at some reputable osteopathic school for at least four terms of not less than five months each, from which he is graduated and receives a diploma. His profession requires that he should be an expert diagnostician, being able to tell the cause of diseases as well as to correctly locate them, and these two carry with them the knowledge of how they should be treated. Massage and the Swedish movement can be practiced by any person with or without experience, or without any previous culture or scholastic training. The Osteopath understands the whys and the wherefores of what he is doing, while the other two need know no more than automatons. One is the result of long study and the knowledge obtained from text-books, and the practical application of the same, while the other two can be given at any time by any person without the requirement of any previous conditions.

Such, in brief, is Osteopathy, stripped of the ambiguities, mysteries and vagaries with which many persons less ignorant than curious have regarded it.

If there is one fact that is more potent and patent in the Bible than another it is this, that in its lids from beginning to end there is no positive instruction given for the use of drugs or medicine in healing or curing the sick. We repeat that among its thousands of injunctions and countless instructions there is not a single passage that can be construed as a positive instruction to use drugs or medicine for healing. This certainly means something. When we consider that the Bible was written by men who believed in medicine and in all material means for healing the sick, we are astounded at the fact that it nowhere teaches the use of medicine. Why was this omitted? Because medicine was not regarded as an essential agent in disease. Can there be any other reason than this? The Bible is the model of all books for plainness and directness of speech. It contains more precepts and examples, more wisdom than all other books combined, and its teachings were intended as lessons to all who would, in an unprejudiced manner, peruse its pages, and yet it has nothing to say in reference to the use of drugs or medicine in the healing art—the art which is to-day so universally practiced in some form among all the nations of the earth.

The art of medicine, like many other arts, had its origin in Egypt, but its early history is entirely legendary. Esculapius became the god of the healing art, and since his day the Esculapians have been deified down to the present time. They have instilled into their followers and imitators, a devotion that has been productive of the bigotry which is their dominant characteristic, and which has stood as a stumbling block in the way of their more rapid advancement.

There is but one passage in all scripture which tells of a physician making a call, or having been sent for. King Asa was the patient. It is written of him that he was diseased in his feet until his disease became very great and he sought a physician. We are not told who the doctor was, or whether he prescribed internal or external remedies. He must have been one of the most learned men in his profession, because his patient was a King and would have no other. But look at the sequel—the result! The good book does not say that Asa grew better or that he alter-

10

nated for days and days between life and death. It is profoundly silent in reference to particulars. The only thing that it does says is: Asa sought a physician "and Asa slept with his fathers." And yet he was not by any means an old man. The only reasonable conclusion that we can draw is, that the King was hastened to his death by the use of drugs, for after the physician came Asa slept with his fathers. This scriptural incident stands out in the light of a solemn warning to those who believe in the power of drugs. The story of Asa is no doubt repeated a thousand times per day throughout the length and breadth of our own country. Many a poor man, as well as rich, is consigned to his grave through the incompetency of those who are called to administer drugs.

God made the herb to grow in the earth, out of which remedies are prepared, but man finds no hint in the Divine Word to justify him in substituting such means for healing. After the redemption of his people from Egypt, the first act of God was to declare himself a healer. Every detail of human life was carefully considered and regulated, but we find no mention of remedies or physicians. If drugs or medicine are endorsed by God, why did not Christ teach their use to his followers, and why did he not use them himself?

"There is a way that seemeth right unto a man, but the end thereof are the ways of death." It might be said that King Asa followed the way that appeared right to him, but as it proved, the end thereof was death certainly to him. We are told that Jesus once used clay and spittle in healing blind eyes. But does any intelligent person think for a moment that the clay and spittle were used as a means to aid Divine Power in opening the eyes? We may rather infer from this very act—"spitting upon clay" —a contempt for material aid.

Jeremiah says: "In vain shalt thou use many medicines and shalt not be cured." Solomon, who was called the wisest man that ever lived, said: "For it was neither herb or mollifying plaster that restored them to health." Isaiah told the people who were with Hezekiah to put a fig on Hezekiah's boil. This is the nearest approach to a drug being used as a remedial agent that we have any account of in the Bible. And still this is not, properly speaking, a drug; neither is it a physic, nor a chemical compound. It was in the nature of an emollient application, consisting of a simple ripe, uncut or unopened fig. This, however, was not the prescription of a doctor, who would have been likely to have applied not only an emollient, but would have given the patient in addition thereto a physic—a cathartic. An Osteopath might, under certain contingencies, give a patient a drink of ice water, or apply a dampened piece of linen to his face, or to any portion of his body; but no reasonable man or woman would hold for an instant that he was using a drug in the common, or well understood acceptation of that term.

Christ and his apostles healed the sick and all manner of diseases, but in no single instance did they resort to drugs. Being inspired men, they understood the laws which govern our physical organism, and the very fact that in the treatment of disease they used the methods that had not been suggested by physicians, and were not in vogue among the practitioners of that day, teach us the important lesson that drugs were not then and are not now to be used as remedies in healing. Christ was both human and divine. His was a master mind comprehending not only the motives of men, but he understood all questions that were entertained or discussed by men of whatever kind or nature. Being often alone in the mountains of Judea, he studied the laws of health, and of being, and

coming as he did to the world as an example to all succeed-
ing ages in everything that tended to the welfare and happiness of the race, he did what he intended to do, and what
he came to do; but in all the history of his life, which was
so replete with precept and example, we find absolutely
nothing that warrants the use of drugs in the art of healing. Had this been a question of no moment, then his
example and the method he employed for healing would
have been of no consequence to us. But the method, carrying with it the power of life and death, and used by
him as an all potent and effective one, shows conclusively
that his preference lay not in the direction of drugs and
medicines.

Then we say that Osteopathy and its teachings are in
full harmony with the Bible. There is no physician, however skilled in the art of healing by drugs, I care not how
good his heart, how great his desire may be to alleviate the
sufferings of his fellow man, who can approach so near the
sick-bed of the patient prepared to help and build him
up as the thoroughly competent and conscientious Osteopath.

There could be no better means employed by the friends
of humanity, means that would more speedily bless the
world, than for the churches to take the Bible in one hand
and Osteopathy in the other, and go forth upon their heav-
enly mission. Could such a thing be, it would not be long
before "the eyes of the blind shall be opened and the ears
of the deaf shall be unstopped. Then shall the lame man
leap as a hart, and the tongue of the dumb sing. For in
the wilderness shall waters break out, and streams in the
desert, and sorrow and sighing shall flee away."

A man owes it to humanity, to the good of the world,
and to his own conscience to surrender his preconceived
opinions when they are shown to be false or not in har-

mony with the enlightened trend of thought. We are not
here to continue any of the old tread-mill processes in
which our fathers or predecessors toiled, not knowing or
believing that there was a better way. Our pride of ances-
try and confidence in the judgment of those in whose foot-
prints we delight to follow, furnishes an incentive for us
to remain in the old paths. Our environments, too, have
much to do in holding us fast to the old beaten track.
But if we are ever to advance; if we ever expect to con-
summate anything for ourselves, we must turn about and
face in another direction. To live as our fathers did does
not satisfy the independent, inquiring mind. We may
live easy and enjoy life. But the opportunity is given us
upon every hand to do more than they did. We possess
advantages that they did not enjoy. The light which
shines upon us has been greatly increased in both brill-
iancy and strength since their day, and if we do not appropriate it to ourselves and help to hold it aloft as a torch
and a more perfect light to others, we live in vain so far
as being thinkers and workers is concerned.

Chirst, we are told, "came to his own and his own received him not." They rejected his teachings. The
Scribes and Pharisees were so deeply wedded to the old
laws and traditions of which they had been the acknowledged expounders for centuries, that any other doctrine or
dogma than theirs was an unpardonable heresy. Christ
paid the penalty of his courage and convictions. The
same doctrine, however, that was enunciated and preached
by him to the multitudes is the same in theory at least
that is received by the multitudes of to-day. The cause
for which he labored and gave his life is still proclaimed
by the Christian pulpits throughout the world. Every new
gospel, therefore, which is totally different in its teachings
to the established customs or order of things is not only

combated by the Scribes and Pharisees of to-day, but its heralds are denounced as fools or charlatans.

The new science of Osteopathy which has so recently shied its castor into the ring has met with its trials and disappointments as all other newly fledged theories do. But being founded upon truth, and inspired with the noblest of all purposes—the lifting up and healing of humanity, it will continue to grow in power and importance until its enemies even will be compelled to acknowledge its merits.

It was said of Lord Beaconsfield, one of the later day premiers of England, that when he attempted to make his first speech in the House of Commons that he made a failure—that through embarrassment he sat down before he had finished his speech. Many of the members of that body were so unkind that they laughed at him. Looking around and seeing this, he became righteously indignant and rose again to his feet and said: "You may laugh at me now, but the time will come when you will laugh with me." His prediction was verified, for no man among the British statesmen commanded greater respect in parliament than Lord Beaconsfield. So it is with Osteopathy. Its enemies may laugh now at its pretentions, but the time will soon come when they will laugh with it.

It asks nothing but the opportunity to demonstrate its virtues; it declares nothing that it cannot prove; it is a message of good will to man, and being in perfect accord with the teachings of the Bible, and having its foundation laid deep in the laws which govern our being, it will, in the providence of God, achieve a universal victory.

Christ's mission upon earth led him out of the old beaten paths that had been blazed by the Scribes and Pharisees, and pursued after them by the common people. He was in the most exalted sense of the word, an iconoclast. He came to proclaim a new gospel; to revolutionize thought

and sentiment, to change old customs; he was the harbinger
of better things, of better ways of living. His new ideas
permeated society, methods of procedure in courts of jus-
tice, governmental questions, comity between rulers,
friendship between neighbors, the economy and conduct of
the home; and in fact there was nothing that was not directly or remotely affected by his teachings. And yet, in
all that he said and did he left no positive instruction to
use drugs in healing.

Although these deductions are inferentially made from
holy writ, could they be rendered stronger if affirmatively
stated? Scriptural authority may be denied; that is, its
binding and moral force upon the consciences of men by
those who oppose the Bible and its teachings. But even
this class of persons, who go so far as to regard the Bible
as a book of fables—casting it all aside as unworthy of
credence, and of no binding force in precept or example,
will think of it as very remarkable that the question of
drugs and their use as a remedy in sickness is nowhere
touched upon in an affirmative manner. Conceding the
Bible to be a book of fables (which we deny) is it not
strange that such an immense volume, which deals with
every theme that concerned man at the time it was written,
should not have given to the world, among its thousands
of myths and fables, something bearing directly upon the
importance of the use of drugs in the art of healing?

But says one: "The use of medicine was so common
and so universal that Christ did not deem it of sufficient
moment to speak of it. Men grew sick and died then as
they do to-day, and the employment of a physician was
just as necessary then as it is to-day."

This is doubtless all true. But it was of the things
that were common among men, their habits, customs, etc.,
that Christ mentioned and dwelt upon. Nothing escaped

his all comprehensive eye. Treatment of diseases by drugs involved life or death, and was then as it is now, of the gravest import. Had it been right or in harmony with either divine or natural law, he would have used them himself, for he was a physician and healed many people. When Peter's wife's mother lay sick with a fever, Luke, the beloved physician, says, that Christ stood over her and took her hand, but gave her no medicine. None of the apostles treated with drugs. Following the instruction by example of their master, they united with him in all things to carry out and promote the new gospel which he had introduced. Old things were done away with, and among these was evidently the practice of medicine by the use of drugs.

But says another: "We cannot get along in sickness without the drug remedies. They are as necessary at times as the food we eat or the air we breathe."

Beg pardon! Have we not shown by the testimony of some of the greatest living witnesses, whose unbiased judgment and superior information cannot be impeached or called in question, that medicine as now administered kills more patients than it cures? (And we have every reason to believe that it is far more fatal in its effects now than it was in the days of Christ.) Can we not show thousands of cases which have been cured by osteopathy without drugs, and hundreds of these given up to die by the M. D.'s, their demise being only a question of a few days or longer? Whoever heard of an Osteopath using drugs or anything as an imitation or substitute for them? and yet look at his phenomenal success; rarely ever losing a patient or failing to effect a cure, when the disease is treated by him in its incipient stages! We do not say it in a boastful spirit; but come and examine the record made by this infirmary; look at the list of names (the roll of honor) and the diseases for which the patients have been treated, and it will be found that a far greater proportion of them have been permanently healed than can be found among a like number of patients that have been treated in the old way. It will also be found that a far greater number of the incurables have been permanently helped than in any other institution where drugs have been administered.

It was thought a few years ago that we could not get along without the lumbering old wagon as one of the modes of conveyance; that we could not get along in the darkness without the tallow-dip or candle; that we could not get along in towns and cities without the old wells or springs; that we could not be comfortable in winter without the old fire-place or heating stove. When Hood wrote "The Song of the Shirt," such an invention as the sewing machine was never dreamed of. A few years ago it was thought that bread and meat were essentially necessary to a man's existence; that the sun dial was all that was necessary by which time could be measured; that a patient suffering with fever must be bled; and that when almost consumed with the same disease a glass of ice-water would be certain death.

Change is written upon everything of a human character and of human origin. But principles are eternal. The great laws which govern and control all matter and spirit are immutable.

GIVE THE OSTEOPATH A CHANCE.

How give him a chance? I will explain. It is strange but nevertheless true, that human nature is so suspicious of any new thing, and especially of any new theory which professes to carry healing upon its wings to those who are bodily or physically distressed, that it must wait and see what there is in the new theory of a remedial character before trying its virtues. Everybody will stand back and wonder what it is, and ask his neighbor if he can explain the mysteries and the methods which this or that new-fangled doctor employs in the treatment of disease.

This condition of wonderment prevails until some acquaintance perhaps—some invalid of years' standing, who has been the patient of numerous physicians and whose stomach has been made the receptacle of a small apothecary's shop of noxious drugs, makes him a call. The patient has become discouraged. His physical system has been so enfeebled and run down that he is but the shadow of his former self. "While there is life there is hope," and although but a spark of life seemingly remains to him, he calls upon the doctor who represents the new theory as a dernier resort. He does not make the call with any fixed expectation that he will be benefited. But having tried everything else, including the nostrums of the different quacks, he goes as a sort of duty that he owes to himself and friends.

He has but little, if any, confidence in what the new doctor does or proposes that will prove to be efficacious in his case, but remembering that he is at the end of his efforts to get well and that his friends are hopeful of some good results, he quietly if not very patiently submits to the treatment. He continues along at first mostly through the persuasion of others but, after a time, he becomes cognizant of the fact that he is really improving. He grows a little impatient, so anxious is he to recover; he never stops to think that a disease which has been fastened upon his system for years cannot be cured in a few treatments, or even in three or six months' time. He finally recovers; the doctor has gained a friend or, perhaps, several of them, besides some helpful reputation. He receives something more of a friendly recognition, and thus with each new patient that he successfully treats he becomes better and more favorably known.

The public doesn't stand quite so far back as it did, and those who stood upon tiptoes at a distance, afraid that their own reputation for good sense would be in jeopardy, come up a little nearer and take a closer view of the man whom the people say has done some wonderful things.

Now, suppose the patient had come to the doctor years ago, when his complaint was beginning to be developed, is it not reasonable to believe that he would have more certainly been cured at that time? When he was younger and stronger and the vital forces of his body were nearer their normal condition?

The doctor deserves far more credit than he receives. He has performed a miracle, not in the biblical sense of the term; but he has snatched a patient from the very edge of the grave, into which he would have gone had it not been for his timely aid. He has restored to life and vigorous manhood a person whose malady was of such a character that it defied the skill of the M. D.'s; a person who could not be successfully treated by them, and who had himself given up all hope of final recovery.

Give the Osteopath a chance!

Don't hang back until you are literally dead, and then

blame him for not resurrecting you. He can do many marvelous things in healing the sick, but he cannot quite bring the dead to life. This is one of the powers which is not possessed by man. Call upon the Osteopath when you are first taken sick and see what he can do. Do not wait until you have become nauseated with drugs, and your ailment has become chronic.

Give him your confidence, and let him see that you do. He will not ask you to take any medicine; he will not cut you up into pieces; he will not gash any portion of your body, nor will he harm a hair of your head. He will handle you gently, intelligently and with a knowledge of your disease that will not only surprise you but which will almost invariably result in healing you. He will not treat you simply for the money that you may pay him. No thoroughbred, conscientious Osteopath will do this. . He loves his profession too deeply to lose sight of it behind the money it may bring him. The fate of Osteopathy is partially in his hands. He lives and works to promote its growth and development, and while he, like mankind in general, may love money for the comforts it brings, he loves Osteopathy more.

To illustrate the truth of what we have said in reference to the character of cases that every new doctor is compelled to treat, before he gains the good will of the community, we publish in this connection the following statement which is only one of many of a similar nature:

"I have suffered from stomach trouble, in its worst form for over seven years. Aside from that, heart trouble and other complicated diseases. For the last six months I have lived in fear of over-eating or over-doing. The least thing caused such weakness and near fainting that life was becoming a burden to me. My bowels for four years have never moved off naturally; have doctored for years, when finally two very able doctors told me to get help I must go to the hospital for an operation, or never would be well again.   I felt very much discouraged.   This was my condition, when a friend of my husband told him of you.   I listened to him with indifference, but was now suffering nearly every day after eating.   Finally I was persuaded to go and see you. I reluctantly consented to take a course of treatment without any hope whatever that I would receive any benefit. Here is the result: after the third treatment I was told to "go home and eat anything you want," but was still afraid, but did venture on a little solid food—it never hurt me. The seventh and eighth treatment the doctor cautioned me to be careful not to over-eat.   I am perfectly astonished at the progress I am making.   My bowels move off naturally. In fact, all the former annoying symptoms are leaving me. I feel very grateful, and can recommend the treatment to anyone afflicted as I was."

This, I say, fairly shows the condition of both the mind and body of the patients who are most likely to fall into the hands of the doctor who advocates a new theory in practice.   He has at first no easy sailing.   His bark is afloat upon the roughest of seas.   He contends with waves, winds, whirlpools, fogs, visible and invisible rocks, and if he finally succeeds in clearing these, and getting into the fair, open sea, he has done well.   He has made a fight that no man could do who was not conscious of the fact that he was battling for a principle.

I will relate another incident that occurred soon after I came to Denver to further show the outcroppings of human nature, and the want of confidence that is generally manifested towards the strange doctor with his recently projected system of therapeutics.

The case is that of a very poor man who did not have

money enough with which to pay his fare on the street
car. He hobbled in upon his crutches, a distance of about
twelve blocks from the infirmary. After being seated, he
told me that he had been suffering for a long time with
rheumatism and had used the prescriptions of many different physicians, but all to no purpose; that he had some
months since lost a good position on account of his complaint, and that he had at length become a burden to his
friends, etc. After he had finished his story, he remarked:
"I have lost all confidence in the doctors, and will tell you
frankly that I have none in you. I can't, however, trouble
my friends much longer, and if I cannot get some relief
I shall do what I have been contemplating doing for some
time." The man was certainly very frank, and his case
seemed to be so desperate that I hesitated a moment about
undertaking it. If I should treat him and fail to do him
any good, I was satisfied that he would speedily carry out
his purpose of relieving his friends of his presence, and
then it might be said that I had contributed to the result.
I finally told him that I would examine him, but that he
must lie on the operating table which was a few feet away.
The instant he arose to his feet he was taken with a pain
so sharp and severe that he would have fallen to the floor
had I not caught him. I gently, but with main strength,
placed him on the table and proceeded with my diagnosis.
He told me that he had been subjected to just such pains,
and they were at times so intense that he could hardly
bear them. I located the seat of the disease in the quad-
ratus lumborum muscle, which has its origin in the crest
of the illium, and transverse process of three lower lumbar
vertebrae which insert into the last rib and is supplied by
lumbar nerves. The man had so twisted himself either by
a fall, or some other accident, that the fifth lumbar vertebra had been sprung, and the innominate bone had been

slightly moved. This produced a strain on the muscles and a pressure on the nerves.

I readily placed the last lumbar vertebra and the innominate bone in their proper positions, and the patient stood up and walked with as much ease, apparently, as he ever did. A month's treatment restored the muscles and ligaments which held the bones to their normal condition.

Now, this man was like hundreds of others who suffer for months and sometimes for years with similar complaints, the attending physician not knowing what is the matter. All the external applications and internal remedies that were ever conceived of would have done him no good. The M. D.'s rely so wholly upon their medicines that they never think of going beyond or outside of these in any case that they may have. They talk about nature in certain instances aiding the patient, but the idea is always coupled with the thought of drugs. They are inseparably associated where nature is mentioned. They think that natural laws, without drugs as a lubricant, are of no force; hence they blindly administer a dose of something that the pharmacist has prepared.

Any skilled Osteopath who has studied closely anatomy, physiology and pathology could have relieved this man and sent him on his way rejoicing, without a resort to either drugs or knife. It was simply a question of knowing how and where to find the seat of the pain, and what to do after this has been ascertained. Some one of the regular physicians might look at a swollen limb, or any other part of the body to see what it indicated, so as to apply his remedy. He never thinks of doing more than this. In fact, his superficial knowledge of anatomy will not warrant him in doing more. The Osteopath, however, so thoroughly understands the entire human mech-

anism that he knows, not only what is the matter, but
the remote or direct causes which have produced the pain.
He may examine not only the immediate locality of the
pain but the muscles, ligaments, or any other portion of
the machinery which bears upon it, to find just where and
what the inharmony is. He may not treat the exact spot
of the pain, but treat some other spot or locality of the
body which produces it. In other words, he gets at the
root of the disease or complaint and applies his skill to
that.

The nutrition, the physical power, the motion and the
repair to the muscles is furnished by the blood; but all
this is governed by the brain circuit.

The blood vessels may be enlarged or clogged, the muscles contracted or flaccid; a bone may be moved or slipped
or twisted, and the nerves inflamed or paralyzed. If any
of these conditions exist, the machine is out of order.
"A fall, a shock, a cold, a strain, drugs, malpractice, bad
or not needed surgery, or physical neglect may cause some
bodily organ to be out of adjustment. The fulcrums
shifted in this mass of interrelated levers, working through,
over and under thousands of nerves, blood vessels, glands
and two hundred bones produce friction or pressure, or
disorganization, or congestion, disease, perhaps death.
The co-agency of the organs is disorganized and their mu-
tual functions interfered with, and this alone is disease.
To illustrate: Twisted, turned or warped ribs cause' asthma, chest diseases, even consumption. False pressure
or friction, strangling a motor nerve produces paralysis;
a sensory nerve pain, the tortures of neuralgia; a symptomatic nerve, disease of the organs of assimilation; an
obstructed artery may cause atrophy or heart failure; an
obstructed vein, catarrhs or rheumatism. The Osteopath
with infinite labor and patience properly adjusts the bones,

normalizes and puts the misfit muscles into their traces; he reduces false pressures; he stimulates or relaxes the great network of nerves that control the functions of every organ of the body; he frees the forces and currents of the human machine until it works perfectly." Health is bound to follow. This is natural. There can be no disease where all functional derangements have been properly adjusted.

In straightening up the human machine, the Osteopath makes no use of the black art, hypnotism, nor any of the fads and frauds which have obtained in the past or present. He does so by his knowledge of the anatomy of the human body, and with the same skill that the knowing mechanic adjusts the machinery which is the work of his hands. It is simply a question of knowing how, and this knowing how does not come from any supernatural agency which the Osteopath evokes, but from his plain, common sense observation of the natural laws by which the human machine is manipulated.

One can never know, who has no experience, the difference between the allopathic and Osteopathic treatments —I mean more especially with reference to the general effects of the two methods. The pill doctor, if you are really sick, or if he thinks that you are going to be, must necessarily resort to medicine which is more or less nauseating, and often gives you stuff at which your stomach rebels. Even the very smell of some of his remedies will disgust and nauseate you to such an extent that the disease with which you are threatened, or with which you are already suffering, will be hastened in its destructive work upon the system. Not only your olfactory nerves and sense of taste rebel against the sickening dose, but nature's laws which have been implanted in your being so lustily cry out against it that the dose is immediately expelled, before it finds a permanent lodgment in the stomach. This process

11

is, perhaps, repeated a number of times and in different forms before the dose is made to stick. In the meantime the patient has been suffering intensely from nausea and vomiting, and has been so weakened and sickened that his vital forces finally surrender. Then he becomes an unwilling captive, and the attending physician congratulates himself upon the success he has achieved in making the dose stick. The poor stomach has been so terrifically assaulted that its natural powers of resistance have given way, and the hero with his pills, pellets and medicaments enters and takes possession of the citadel of life.

This is simply one of the pictures drawn from life, as witnessed almost daily in the sick room.

Turn to the other picture—that of the Osteopath treating a patient. It makes no difference what is the nature of the disease; however delicately organized may be the machinery of the body; however sensitive may be the nerves of taste or smell, these are not to be touched or tempted, either by insidious approaches or open attack. No medicines are given either externally or interiorly. Nothing is done to produce either mental or physical exhaustion. The building up methods of the Osteopath are helpful, wholesome, strengthening and pleasurable, and these results follow not only every treatment that is given, but they abide with the patient who is faithful to the instructions of the doctor. The Osteopath aids nature; the allopath with his pernicious drugs (and all drugs are pernicious) weakens and retards nature in her recuperative work. Which is the more natural, and if the more natural, the more beneficial treatment? What does nature dictate? What does the revolt of the stomach teach? What do reason and common sense say? What do ease and comfort suggest? Is there any fair-minded, unprejudiced man or woman who would not say that Osteopathy was as much preferable in its treatment of a patient, to the methods of allopathy, as the comforts and conveniences of a modern home are preferable to the comforts and conveniences of the pioneer log cabin?

Yes! Give the Osteopath a chance.

But do not give him the last chance. Do not slight him in this way. He may save you weeks, months or years of ill-health, and in doing so he will save you perhaps much precious time and money, which are becoming more important as the age advances. Go to him, as you have been in the habit of going to your family physician, when you are first taken sick, or when you meet with an accident. Don't wait to try somebody else first. If you do, you may regret it. If the Osteopath can do you no good, he will undoubtedly do you no harm. He will not reduce your system by giving you unwholesome or baneful drugs; if he cannot build you up, and make you whole, he will not tear you down and weaken your vital forces. You will find him to be a gentleman in the broadest sense of the term, and not totally ignorant of the amenities of good society. You will not find him to be extravagant in dress, nor using very bad grammar, and yet his success does not depend on the observance of the small conventionalities of life. The best work in the world is done by the ordinary people.

With every true Osteopath, the considerations which weigh more with him than all else outside of his love for his profession are those of character and personality.

"His practice brings him into direct, personal relation with people. Indeed, his treatments are strictly personal —upon the very persons of his patients. It is by the touch of his hands that his results are obtained. This calls for the courtesy and chivalry of the best type of appreciation and refinedness. Rudeness, incivility, familiarity, intimacy,

immodesty, whether of speech or action, will detract from
and destroy the fineness of effects which a true personality should create. Carelessness, inconsideration, or in-
appreciation of the condition, or circumstances of a patient,
would always be a mistake. Candor, honesty, sincerity,
and truthfulness, should be shown in every case. It is best
to show these traits as a matter of principle, and it will
pay to practice them. To promise results that are not
warranted, or that are not readily expected or worked for,
in a case, for the money there may be in it, is fraudulent.
To take advantage of the hope of health, by a promise
that cannot be fulfilled, is the crime of cheating. To use
the money that is justly earned in hard work, in a reckless
or pompous way, will be taken as a sign of weakness somewhere. In all these things the Osteopath can win golden
opinions to himself by being wise, prudent, thoughtful,
and considerate. The code of ethics for Osteopathic practice has yet to be written. Such a formal code may not be
necessary. If the Osteopath will be guided and governed,
in his person and in his profession, by maintaining the
very best of character, by practicing the very best of common sense, and by showing the very best of culture."

One of the most remarkable cases brought to my observation was the following:

A little boy seven years of age was brought to the infirmary to be treated for epilepsy. He was not only very
frail, but was most ghastly in appearance. There did not
seem to be as much life in him as is to be found in an
ordinary healthy babe. The child had suffered long, and
the sufferings were the result of ignorance. The disease
was produced at its birth by the physician in charge of the
case—by twisting the neck and causing a strained condition
of the muscles and ligaments so as to dislocate the clavicle,
and cause pressure on the phrenic nerve which supplies

the diaphragm; the pneumogastric nerve was also affected, and the circulation to the head was irregular. Upon the least fright, excitement or exertion, the child would hold his breath until he became black in the face—sometimes until he lost consciousness. After regaining consciousness, he would be so limp that it apparently seemed that all life had left him. His appetite was bad—would eat no food that had to be masticated. His diet was principally milk, and this had been the case since birth. His bowels did not move only when injections were used, and even then did not move with freedom. He was also troubled with sleeplessness, and would wake up during the night and have one of his spells. The doctor who pronounced the disease epilepsy advised the mother to take the child to a lower altitude. She went to California and remained two years, but the patient grew worse and the physician there told her that he would continue to grow worse as he grew older, and that there was no cure for him. The mother, upon her return, came to our infirmary. After we had treated the boy about a month, all symptoms of epilepsy left him, and now for three months not a symptom of his disease has reappeared. He eats heartily of anything he desires, sleeps well, and has free and perfectly natural movements of the bowels; he is a smart, active, rosy-cheeked boy.

This next case was of four years' standing, brought on by la grippe—caused rapid pulse, sour stomach, pain in the stomach, fainting sensations, pain in chest, hot and cold feet, pain in left side, which was always aggravated after movement of the arms; bowels moved 15 to 20 times during the day, and at night patient was feverish and sleepless, and generally exceedingly nervous. Was also greatly annoyed with catarrh, which manifested itself through the nostrils and in the ear, the latter organ having become deaf. The patient had been treated by numerous physicians, and of course had taken any quantity of drugs, but to no effect. She came to the infirmary and after about four weeks' treatment her disease had almost entirely disappeared, together with its complications. Her bowels are in a normal condition; the running at the nose and ear has ceased, appetite is good, and the patient says she is absolutely well.

The next was a case of sick headache, constipation and stomach trouble. The patient is a gentleman well-known in Denver and Colorado for his ability and attainments. He would be greatly distressed after eating the most common meal, and feeling a spell of headache coming on, would lie down to sleep if possible. He was subjected to these spells when off on journeys, and for lack of sleep and suitable accommodations he would at times suffer most intensely, being often confined to his bed for days. During such periods he could get no relief only through sleep or movement of the bowels.

A diagnosis of his case showed that the first and second cervicle vertebræ were slightly displaced and twisted, causing a contracted condition of the muscles, and the ligaments stretched, which caused an irregular supply of blood through the vertebræ artery, causing pressure on the brain, and a lack of oxygen through the brain. There was also a pressure on the gall-duct, and a shutting off of the bile from the intestines, causing inaction of the bowels for want of lubrication.

In the treatment, the first and second cervicle and the gall.duct were adjusted. The patient was treated for two months, and regained his health entirely. Of course, he, like the other two mentioned above, had tried many of the "regular" physicians, but could find relief only in Osteopathy.

A patient came from New Mexico, greatly afflicted. He could not lift the arm from his side. Circulation was impaired; he suffered with constipation and stomach trouble, dizziness, sleeplessness, paralysis and dislocation of shoulder. The pain in his shoulder was very severe and would often continue two or three weeks. The patient had two years before fallen from a horse and had been dragged by the animal a distance of 20 or 30 feet. Had pain in right breast, and under the right shoulder blade. Had also a contraction of one side of the face and jerking pains through the body. Had an intense burning sensation in his head, which was caused by the right clavicle being slightly turned and jammed in irregularly at the sternum and articulation of the scapula, and clavicle under the point of articulation with the scapula.

The second and third cervicle posterior and the left auricular nerves were also affected, causing pain in head; also a tightening of the levator anguli scapula muscle.

After the second treatment, the pain had left the arm and that limb was more supple; pain had left also the head; paralysis disappeared; muscles of the arm greatly relaxed and not painful. After the fourth treatment, the patient could raise the arm over the head with ease; could not raise it at first at all; his bowels are regular, and he has a good appetite. The patient in speaking of his case said: "This is the first beneficial treatment I have had. Have had three physicians during the past year; a specialist, homeopathic and allopathic; they did me no good, but great injury by filling my system with poisonous drugs; the drugs relieved me only for the time being, then I would grow worse."

The doctors had been treating him for heart trouble which he did not have, but which they said was caused by falling from a horse; they said that an artery would eventually burst in his head, which would cause paralysis and then be succeeded by death, etc., etc. The patient was treated three months and is as well now as he ever was.

DOCTOR SQUINTER.

The following poem so admirably, so facetiously and so squarely hits the nail on the head that we give it space. The "Doctor Squinters" are by no means uncommon figures upon the professional stage, and may be seen in the older settled neighborhoods of every state in the Union, where they pursue the even tenor of their way, never dreaming that the closing years of the present century sound the death knell of many of the superannuated thoughts and methods which now obtain in the healing art:

DOCTOR SQUINTER.

BY N. J. SCURLOCK.

Old Doctor Squinter knew the ways
By which his kind wins cash and praise;
He took none of his drugs himself,
So lived to corner fame and pelf.
His dear diploma, handed out
As license on the road of doubt,
Well served him as a legal shield,
When censure's shafts flew o'er the field.
Of stoic turn and pompous mein,
He fattened while each purse grew lean,
That helped to pay the heavy price
He put on powders and advice.
A rank empiric, through and through,
He ridiculed departures new,
As cut-and-tries from Galen down
Have done in quest of cheap renown.
He was one of that bigot class
That once swore Harvey was an ass,
And proved by all the narrow school
That Jenner was a vulgar fool.

The oldest methods, oldest drugs,
When death he harnessed, served for tugs,
And hard he made the "pale horse" work,
By aid of his prescription clerk.
His good luck walked with some around,
His errors lay safe under ground.
In his sham battle with man's ills
He used for bullets countless pills,
And, let the end be death or cure,
Of his bills only he was sure.
By belladonna still he swore,
And leagued his faith with hellebore;
The weaker any patient grew,
The more on strongest drugs he drew;
In fever he used calomel,
And if it proved typhoid—ah, well!
The undertaker was his friend,
The tombstone dealer would attend.
At homeopathy he raved,
And vowed the whole world was depraved,
When any puny dupe or fool
Would trust his carcass to that school.
By dogmatism's vain pose enticed,
He scoffed the healing done by Christ,
And claimed the lowly Nazarene
Ne'er dreamed of cases he had seen,
And back to former health had led,
Though death was crouching near the bed.
So did he practice, rant and boast,
While patients yielded up the ghost,
Or dearly paid to live, and learn
The use of drugs that blight and burn.
He tried and guessed, and guessed and tried,
Until the sick got up or died;
He laughed with good luck, cried with ill—
But charged for either in his bill.
So long the use of drugs he knew,
A monomaniac he grew,
And just one instance we relate
To show his pitiable state.

Old Doctor Squinter always sought
A chance to air superior thought,
And one day as he drove along,
Saw in a field an idle throng—
The owner "green," the thresher new—
No fellow there knew what to do.
The skies were dull with threatened rain,
The farmer feared 'twould catch his grain;
But something ailed the new machine,
Though what it was could not be seen.
Old Squinter pulled his horse up—"Whoa!
Say, oil her, boys, and she will go;
I doctor bodies or machines,
And know what oil in plenty means!"
Both oil and power the men turned on,
The wheels confessed their cunning gone.
Just then a neighbor reached the scene,
Who once had run a like machine,
He looked, and laughed, adjusted then
Two belts, and called out, "Ready, men!"
The thresher right began to hum,
While Squinter only growled, "By gum!
That [1]Still I hear so much about
Would say that proves his plan, no doubt,
But all such argument's too thin—
The oil has now just got soaked in!"

[1](Note) Dr. Andrew Taylor Still.

A SKEPTIC INVESTIGATING OSTEOPATHY.

From the Journal of Osteopathy.

To show some of the ideas entertained by even intelligent men, in reference to Osteopathy, we here reproduce an article, entitled "A Skeptic Investigating Osteopathy." It is from the pen of a wide-awake newspaper man, and will be read with interest, no doubt, by persons whose ideas of the new science are about as imperfect as his were before he began investigating it:

"Newspaper men are seldom among the early disciples of the world's real performers. They understand somewhat society's penchant for being gulled, and it makes them overly suspicious of the man who rises up to preach any new gospel—especially if its acceptance means any vast credit to the preacher. They see so much of fad among the people who feed on notoriety, and so much fake by people seeking financial aggrandizement, and such stupendous folly by hair-brained inventors and discoverers that they are loth proverbially to grant to any new Prometheus the credit of bringing down intellectual fire to his fellows until it has been set out and proved its right to recognition by kindling a mighty conflagration.

"This conservatism, however, does not apologize for the newspaper man who fails to know just the moment that the public mind adopts a new thought into the kinship of public opinion.

"I laid off the cloak of skepticism on a recent visit to Kirksville, and to my own amazement learned of the birth of a new science—a science which I could understand, and see demonstrated, and speculate upon as we do the facts and principles of chemistry and astronomy. Perhaps the way a newspaper man was overwhelmed and forced to capitulate in spite of himself may have interest for the public while focusing attention on this field of health suddenly brightened with a new search-light. It is a field where the lines of battle are now forming that must wage unending conflict for a generation. The issue of this battle is of more consequence than wars for the independence of states or the possession of territory. The stake is man's right to live.

"In the course of duty I was sent by the St. Louis Chronicle to visit the fane of Osteopathy. I had heard of this school of doctors as a bone-setting coterie which tied its faith to bone carpentering. Casual inquiry in St. Louis gave me this additional explanation:

" 'Bone doctors trace all ills to broken and dislocated bones; they pretend to cure everything by twisting and resetting bones, and if a diagnosis fails to show bones out of place a bone will be broken somewhere—preferably a small one in the foot—and it will be set again.'

"Early one bright May morning I set out in Kirksville to feel the village pulse concerning the local prophet. I have great faith in the ability of neighbors to know neighborly weaknesses—particularly impositions by which the other fellow is making money. None was found to discredit Dr. Still. Not one even shrugged his shoulders when discussing his system. It was short work to find out that the people who had dwelt nearest the Missouri sage believed in him. They respect him and love him.

"The hotel man was the first to tell me that Dr. Still was a benefactor of his fellows. It was patent this might apply with special force to the hotel business if Dr. Still drew strangers to the village.

"The depot master and telegraph operator told of scores

who had come on beds and in stocks and had departed
with spirits and step elastic. Recollections of an assembly
of faith healers, and again of standing in the outskirts of
a crowd harangued by Indian root venders, and hearing the
multitude itself proclaim its own healing by means of faith
specialists and fake root bitters, came to me. There was no
proof for Osteopathy even in a multitude of witnesses.

"Merchants and bankers were as eager to add their en-
dorsements. These seemed to be prosperous. The streets
in front of their business places showed life and trade.
Health was admitted to be the chief resource of the village.
Clearly, then, bankers and merchants might look at 'healing without drugs' as a good thing for Kirksville, whether
it helped others or not.

"Like the chemist, every reporter has special tests in
emergencies upon which he relies to throw down precipi-
tates that show the constituent part of things. I went to
the doctors whom I had no right to suppose were subsidized
in any way by the system which cuts off their living. They
spoke a good word for Osteopathy. In wonder I talked
with druggists. These said Dr. Still cured without medicines. I took a final appeal to the undertaker. He said his
line was depressingly dull, despite the influx of invalids in
Kirksville, because 'the old doctor' usually put them on
their feet again. My test-acid of conflicting interests in
this case had failed me.

"What wonderful unanimity; what uniform loyalty to
home institutions, I wondered, on my way to the infirmary.
Yet, there was no doubting further that a whole lot of people believed in Osteopathy, and at least one prophet had
found fame beside his own vine and fig tree.

"My assignment became suddenly more interesting and
I set out to discover what sort of a rabbit's foot the head of
this new sect used in his business. I have always been

willing to receive pointers on driving the multitude to believe what you tell them.

"Patients are thick in Kirksville. So are the disciples who sit at Dr. Still's feet learning the strange truths he teaches. I met both orders at every corner. They talked freely, many intelligently, some scientifically, upon their treatment and studies. All were interesting.

"It proved easy of definition and demonstration.

"Osteopathy is a science of restoring health to body and mind by the mechanical processes having to do with forces inherent in the body and independent of drugs, except for antiseptics and antidotes for poison. It reveals that the body has the power and the appliances within itself to remove disease just as easily as to produce disease, health being the harmonious and normal working of all functions, and ill-health being a condition which may be normal although inharmonious. It demonstrates that this is true by ascertaining what are the functions of bone, muscle, nerves, blood vessels, glands and juices. It shows that it is more than a theory by going to the facts of the body and by manipulating them to produce desired consequences. Although its principles are plausible and fascinating, its facts are astounding, for it is as yet a science of facts, of observations, of demonstrations, with few theorums which can be offered as authoritative. Its theories and philosophy may be furnished to the world in surprising completeness by its author before he dies. They may become his legacy to his generation in posthumous papers. They may have to be evolved largely by his disciples.

"It gives the sick confidence to be told that Osteopathy demands the most intricate and exact knowledge of physiology, anatomy, symptomatology and pathology, and the graduate of the American School of Osteopathy must have a far more intimate knowledge of the human body than

surgeons usually acquire and which physicians as a rule
do not dream of. Upon this knowledge, with the principles applied, are wrought results—health.

"The patients hailed from everywhere and some knew
a lot about medicine, surgery and sickness. They had
been investigating from necessity in many quarters, seeking relief. Their stories were almost incredible. The
blind had come to see. The halt walked. Epilepsy had
been banished with the simple readjustment of a bone out
of position that paralyzed some functional nerve or artery.
One patient, who had been brought from an insane asylum,
was endowed with sense in a few weeks, and he straightway
enrolled as a student and began to prepare himself to practice. Goiters galore and gigantic had been removed without pain. Nervous prostrations till patients could not rest
were cured, and the whole catalogue of woman's ills had
been banished for all except those who still put their trust
in old time medicines.

"Fevers are aborted in a few hours, and all other acute
disorders, I am told, yield with mathematical precision to
this new method, with no other outside agency than the
intelligent direction of the forces within the body itself.
None could ask for greater evidence of the sort that indicates everything but proves nothing. I resorted to those
capable of teaching to ask for scientific and rational reasoning. I as yet had no adequate conception of what is meant
by Osteopathy. Was it another blind reach after mystery,
appealing to isolated facts to uphold blinder theories? I
should never credit it with a mountain of miracle behind
it, unless it appealed to my sense as the most sensible sort
of thing, and could be demonstrated in reason and felt in
the hand as tangible fact.

"Faith does not enter into the science any more than
physics. Physical laws assert themselves with each test

whether experimenters believe in them or not. When one ponders on the experiment of putting strong, noxious drugs into the stomach and blood to upset all normal conditions, and realizes that no man can tell what the harvest will be, it is not to be wondered that the patient with fever will prefer the ministrations of those wonderful Osteopathic fingers, with their miracles of potency, while the mind can be regaled with such lucid explanations of how disease came in, and how it must be driven out.

"How far the new vision of sickness will supersede the administration of strong drugs, or drugs at all, in the course of the next generation no one can say certainly. Yet some things are certain. Diseases will be cured, which could not be cured by drugs and the knife. Surgery will find itself doubly efficacious with a knowledge of Osteopathy. The use of "horse" medicines will be forgotten. Probably instead of fighting the inevitable, medical colleges at the demand of surgery will incorporate a part of Osteopathy into their curricula. Physicians who do not want to go back to school to learn will first scout Osteopathy, and when it prevails in spite of them will claim it is but one corner of the field which they early explored and have always used in practice. It is my belief that a school of practitioners will arise who will make use of the new science in concert with a limited use of milder medicines, such as the homeopaths have made popular. But the founder of the new science believes that the art of Osteopathic practice will render the further use of drugs unnecessary.

"Who can fathom the possibilities which Andrew Taylor Still may have brought to his generation? I regard the study of Osteopathy as the most alluring field which any young man or woman might enter, whether the rewards sought are attainments, good to one's fellowman, or hard silver dollars. The generation of bright, well-equipped college graduates who are pondering upon how they can make a living in hard times and perhaps win fame, cannot afford to overlook Osteopathy. It holds laurels for the student, particularly with a biological education, and for the practitioner not equaled, in my judgment, in any other field on earth. Osteopathy is the opportunity of an epoch!"

It will be noticed that the writer says, that in his belief "a new school of practitioners will arise who will make use of the new science in concert with a limited use of milder medicines such as the homeopaths have made popular."

This may come to pass. We do not know just what will occur in the future. Of one thing, however, we feel quite sure. No true Osteopath would ever be found in a scheme of this kind, for this would be a partial compromise with drugs. Osteopathy has put its hands to the plow and can never turn back. The further the new science is tested, the more deeply convinced are its followers that it is all sufficient within itself, to stand alone and independent of all other theories which have drugs for their groundwork. What other schools of medicine may do towards the employment of Osteopathic methods, we cannot tell. When they fully realize that Osteopathy has become an undoubted success, and that humanity calls for it as the only natural remedy in healing we may look for changes which will embrace not only a compromise but an entire surrender upon the part of some other schools. We look for a dreadful shaking up, and a turning over of the dry bones, which have so long lain in the way of progress in medical science, and when that time occurs, towards which the adumbrations of the present unmistakably point, then Osteopathy will occupy its own niche, and command an influence which will be worthy of its mission to man.

12

Unlike all religious cults and systems of medicine, many of which have been in existence for hundreds of years, Osteopathy, per se, is not an offshoot. It has neither homogenous nor consanguineous connections with any systems that have preceded it. It is distinctively sui generis. Its oneness and characteristics are its own, and never can be confounded with or merged into anything else. Owing then, nothing to other forms of practice, it will continue to live and thrive as an all potent, unique truth.

LESSONS WE LEARN FROM ANIMALS.

The laws of nature are instinctively followed by the animal creation lower in the scale of being than man. They unwittingly live them throughout the entire period of their existence. They are not personally, that is consciously, responsible or amenable for the infraction of any of the laws which govern their being, and yet they suffer in common with man for each infringement they involuntarily commit. Unlike man, however, they use no material aid or remedies other than those suggested by nature or an inward impulse. They have diseases, undergo the most serious injuries, and not infrequently are subjected to violent epidemics, and yet recover without artificial aid. This, of course, cannot be denied, for observation all along the pathway of our lives establishes it as a veritable fact.

If the animals recover their health and normal conditions without drugs, why should not man? Is there any logical reason why a man should resort to drugs and the animals should not? They are both endowed with very similar organisms so far as their bodies are concerned; they have similar powers of digestion, absorption, etc., and these functions perform their duty as perfectly and as wonderfully as they do in man.

A sick or injured animal, even the most intelligent, will never take a dose of medicine without having it in some way forced upon him. His instinct teaches him that it is unnatural, and furthermore that it is something that will be hurtful in its effects. His propensity says "No! I do not need it." His refusal to take it is the argument which nature has given him with which to appeal to our reason. The animal's instinct so far as his bodily ailments are

concerned, and the way in which they should be treated,
is safer and wiser as a guide to him than we are to ourselves
with all our boasted wisdom.

Men and animals are both creatures of habit.   The
former become so by indulging their desires in any given
direction, and the latter by involuntary inclination.   The
former drifts into certain channels consciously, and the
latter without realizing at any steps the conditions which
impelled them to act.   In being subjected to these habits,
it will be observed, however, that the animal without being
driven into absolute straits of destiny, never oversteps the
boundary of the natural laws.   They are kept intact more
closely than they could be by man, for the reason that
the animal has no desires which are the result of reason
and reflection, that are to be gratified.

Certain animals will eat ordinarily only certain plants
or foliage in the vegetable kingdom.   Certain other animals
will eat certain other plants, and will not cross the line
which nature has established, instinct teaching them that
it is disease or death to venture beyond.   Man partakes of
any and everything in both the animal and vegetable
worlds that his appetite craves, and yet with all his fore-
thought and experience, he does not succeed in warding off
sickness.   Man when afflicted sends for a physician.   The
animal is cured alone by nature.   If one recovers without
the aid of drugs, why should not the other?   .

But go a step higher in the scale of being.   Among savage nations no real medical treatment is employed.   They
have their medicine men (so called), but they suppose
themselves to be charmers.   They depend upon incanta-
tory formulas and meaningless ceremonials to frighten
away the evil spirit with which one is subjected to when
attacked by sickness.   These savage nations were as plain
and simple in their habits as the animals they chased, and,

like the animals, were seldom sick. When America was discovered the Indians occupied the entire continent, or rather more correctly speaking, their tribes were scattered over every portion of the domain from which they could draw subsistence from forests and streams. They were numerous. Some were a little more enlightened than others, but all were savages in their habits, customs and manner of living. Their decrease in population began upon the advent of civilization. The white man brought with him not only his "medicine men," but new methods of living and new ways of enjoying life. The savages were taught by precept and example what they supposed to be higher ways of life. These were gradually seized upon and followed with avidity, especially the vices which characterized the civilization of that age.

They began to wane in influence and to decrease in number; not even the potency of the white man's drugs could arrest the diminishing process. As long as they were in a savage state they multiplied and replenished. The charmer with his magic was more efficient in saving life than the doctor with his genuine drugs. The savage lived and multiplied in the absence of drugs. He died like all other human beings it is true, but lived out his allotted time—three score years and ten, just the same.

But ascend another step.

Sir John Forbes, physician to the Queen of England, says, that among half civilized nations that medical treatment is either not employed, or consists in mere charms, such as obtain among savages, and yet, they, like the Indians in their barbarous condition, have a less number of deaths in proportion to population than the most enlightened nations.

In addition to these facts there are thousands of instances, even in civilized countries here in America, for example, where persons who have been seriously sick were unable to procure medicine or the services of a physician, and have gotten well. Even in our own state of Colorado, during the earliest days of its settlement when the few pioneers were scattered and living in remote regions, instances are related of cases of sickness where neither medicine nor physician could be procured, and yet the persons sick recovered. The pioneers themselves were rugged and enjoyed, generally speaking, better health when they lived at an inconvenient distance from a doctor. This means that the less they had to do with drugs the better off they were so far as the health of the body is concerned.

What does all this prove? What is the only reasonable conclusion to be drawn from the facts stated? That they are facts no well-read man can dispute. They prove that nature is the real healing power, and that nothing artificially employed can be substituted for or contravene nature. It proves also that when the laws of nature are impinged upon that the results are far more disastrous than they are or have otherwise been.

But it may be asked: "If nature restores the animals, the savage and half civilized races to health without drugs, what is the use of an Osteopath?"

The work of the Osteopath comes in just here. He assists nature, as the engineer assists the clogged or impaired machine. He unlooses the machinery and readjusts it so that it may move naturally or as it was intended to be moved. Osteopathy tones up the system and stimulates the circulation so that the disorder, whatever it may be, disappears. Osteopathy does not make the cure but helps nature, which is the great physician. Nature must repair every break and rebuild every tissue, and often does this in spite of the pernicious influence of drugs. Osteopathy introduces no foreign substance into the body, but by perfectly rational methods the human machine is put in such a condition that each part will do the work that the Creator intended it should do.

Men have, in all ages of the world, sailed every sea and visited every clime to find the elixir of life, or to find that which would restore them to health. They have done this only after they have tried the medicines of the best and most noted physicians in their own, as well as in other countries. The most (supposed so at least) efficacious remedies having failed, they seek the virtues of other climates and try nature to see what effect may be obtained without the use of drugs. Ponce de Leon thought he might find the elixir of life in Florida; Sir Walter Scott in Germany; Robert Louis Stevenson on the island of Samoa, and others in different portions of the world. They had used all the man-made remedies that had been suggested for their diseases, and the thought finally came that they had better try climate and leave all to nature. But they were too tardy in their decision. Nature's laws had been so broken down by long years of continuous use of drugs that they had lost their recuperative energies, and the distinguished patients went the way of all the earth. This teaches us the lesson that we should not look to nature as the last resort. She is a kindly mother and will do all that she can for her children, but like the human mother, there is a boundary to her efforts. In her efforts to right and adjust matters, she is eventually deprived of her powers to do good from sheer exhaustion and overwork. Do not fatigue or wear her out. Solicit her life-giving and friendly offices, more magical in their potency than any nepenthe of either ancient or modern times. Court her smiles and her favor, for she will clothe you in attire more regal than that ever bestowed by the most kingly druggist, and crown you with blessings more price-

less than the gems of Golconda. She will clothe you with
beauty, strength and health and crown you with a robust
and happy old age. She offers you immunity from disease,
an alchemy that will prolong life. Do not reject her offers,
but accept them now. "Procrastination," we are told, "is
the thief of time." It may be the thief that will rob you
of your health, and with the loss of health your happiness,
and when happiness is gone what avails life?

Climate may and does undoubtedly do much towards the
building up of the system. The climate of Colorado is
especially productive of the best results that can be attained by nature. Every sensible physician will tell his
patient that the most effectual remedy for all diseases is
nature, and to spend as much time as possible in the open
air. Colorado is preferred as a health resort to every
other mountain region, because the climate is such that life
can be spent nearly all the year round in the open air.
There are upon an average, during the year, but about
sixty days when the weather is below freezing point. The
days are generally perfect and full of sunshine, and every
physician knows the importance of sunshine. There are
no depressing heats in summer, no sleepless nights and no
continuous rains, and the atmosphere is so clear that mountains thirty and fifty miles away seem to be but ten miles
distant. Had Robert Louis Stevenson tried the climatic
effects of Colorado and relied upon nature for that assistance which he failed to receive at the hands of learned
physicians, his life might have at least been prolonged.

WOMAN IN THE NEW SCIENCE.

In the great advancement which is being made in Osteopathy woman will be included among its teachers and devotees. In fact, she is now taking hold of the science with as much interest and determination to carry its blessings to humanity as the most enthusiastic among the sterner sex. Why not, as so much depends upon the delicate touch of the hand? And whose hand is softer or more sensitive or more gentle than woman's? And who has more heart than woman? Who more of a sympathetic, intuitive nature than she? Her intense love for the helpless, the halt, the lame and those who suffer from the pangs of disease will make her an important adjunct or acquisition in the ranks of the true osteopathists.

Many of the barriers and prejudices which have handicapped and hedged her about have been swept away by the advance of modern thought and civilization, and she now stands in avenues which lead in all the directions which have hitherto been traveled by men only, as a general rule. These avenues are now open to her, and the goal at the end is as inviting to her as it ever was to her liege lord. Although the present is distinctively woman's age compared to what the conditions were which surrounded her even three decades back, the next two or three decades will reap more largely the beneficent fruits of her wonderful development which will spring from her still greater freedom.

Numbers of brave women are now identifying themselves with Osteopathy and are doing as much to demonstrate its truths as man. It is a field wherein there can be no obstacles to her complete success. When her

knowledge of anatomy and physiology is supplemented
with her intuition and the natural glow and warmth of her
heart, her efforts will not only be rewarded with a harvest
of shining dollars, but with the consciousness of having
done something for the betterment of mankind. She has
in Osteopathy a broader field in which to operate for her
knowledge of the new science, while not curtailing or
abridging any of her desires or opportunities for doing
good will bring her into closer relations and into deeper
sympathy with her fellow men; into deeper sympathy because she will then be able to interpret more intelligently
the secret springs of the causes which produce pain and
suffering.   This new science will place in her hands the.
instrument, so to speak, wherewith she may the more effectually labor for the good of mankind. She then becomes in a two-fold sense the ministering angel, combining
with charity the power to absolutely alleviate human misery.   Her touch will be magnetic; her presence will be in-
spiring, and with every condition, qualification and equipment in her favor she is bound to succeed. Then, too,
the field is a new one. But a few nooks and corners of
its broad area have as yet been exploited, and entering as
she does its recently discovered domain, she will commence
with others at the initial point, and in common with those
who have chosen the practice of Osteopathy as a life-work,
she will enter in and take possession of a land that flows
with milk and honey. Entering as she does at this time,
at the beginning, she will escape much of the criticism and
unmanly jealousy which characterized the M. D.'s when she
dared to study their profession and become one of them.
They thought, at least the most ancient and selfish among
them, that she was trenching upon their sacred and exclusive rights; rights which they had alone exercised
through the ages, and thinking that she was out of her

sphere entirely, her claims as a physician were not fully
recognized as compared to the claims of the sterner sex,
who were fortunate enough to be called "Doctor." Nor is
this prejudice dead even to-day, for it still stealthily lurks
in the silent sneers of many of these old moss-backs, whose
ideas of the professional woman, could they be intelligently
expressed, would not be at all complimentary.

In the new field she receives a cordial welcome, where
her ability, her sincerity and her efficiency will be recognized and appreciated. In the ranks of the Osteopaths she
will aid in spreading this new science and be largely in-
strumental in bestowing its blessings upon suffering humanity. Equipped not with the knife of the surgeon or
the anodynes of the "pill doctor," but with a thorough
knowledge of physiology and the anatomical construction
of the human frame, supplemented with her gentleness
and her woman's intuitions, she will make a rich conquest
of grateful hearts. She will go forth conscious of the fact
that her own is coming to her and that she is but seizing
hold of her birthright. All things in osteopathy are hers,
and all she needs to do is to gird on the armor of courage
and confidence and earnestly and honestly seek them.
The goal to her is as tempting as it is to man, and although
she may not be so tall in stature and stalwart of limb, nature has placed in her more delicate hand a wand, the legitimate use of which will draw to her the prize for which
she labors. The pathway may not at all times be strewn
with flowers, but the power of all-conquering love with
which her woman's heart is filled will so smoothe down its
angles and corners that she will make it a royal highway,
wherein suffering humanity will delight to travel.

Progress is the watchword of the day and hour, and
no science has so wonderfully demonstrated this fact as
Osteopathy. From the little obscure bantling of a few

years ago the child has grown and thriven until it has a name and a fixed habitation. It is being rapidly recognized as one of the most promising youths of the country, and ere long the boy will become the grown-up man. Let woman unite her efforts with those of men in the culture and training which are yet to be bestowed upon the aspiring "youngster" and the two forces will give it such momentum that greater results will be achieved during the next few years than it is possible to now anticipate. Beauty, strength and solidity of character are the prominent features which are being developed, and these are expanding and rounding out year after year so that the near future will behold one of the most interesting figures of modern times. It is the special province of woman, endowed as she is for the work, to aid in this wonderful growth and development. Osteopathy claims her not as an adjunct, but as one of the chief builders in the temple whose foundation has been so successfully laid.

Apropos to the subject of woman in Osteopathy, we take the following from the Journal of Osteopathy in its issue of last June:

"The present demand for well drilled osteopaths greatly exceeds the supply. Letters are received by the secretary almost daily asking that operators be sent out into the world. These requests come from communities where Osteopathy has been known by its work. But every competent osteopath is now pleasantly located and overrun with work, while the total number of students now in the school would not, if graduated, supply the demand from the State of Missouri alone. This demand for osteopaths will increase. Every day cures are being accomplished at the infirmary, and these people go home and tell their friends about the new method. Thus the field is broadened and new communities where Osteopaths could step into a good practice are daily added to the list. Young men and women who are about to choose a life work should investigate Osteopathy by all means before casting their lot. There is no profession in which youth and brains will find a more pleasant or profitable employment.

"There is no avocation in life which places within the reach of the industrious young men and women of to-day as great opportunities as are offered in the science of Osteopathy. Other trades and professions are full to overflowing; many are so badly overdone as to be unremunerative to even their most experienced and competent followers. Osteopathy is new. Its absolute success in dealing with disease is a guarantee that the young men and women who equip themselves with a knowledge of this new philosophy will reap a rich reward in worldly goods, and, what is greater still, will 'live to bless mankind.' The world is full of disease and suffering which all other systems have failed to benefit. The practitioner who can reach these people and give them relief will find the public ready to give him a generous reception."

WHO SHOULD PRACTICE OSTEOPATHY?

This is a question of vast moment to the friends of the new science, for upon its correct decision largely depends its success or failure. If the science were further advanced in age and well established throughout the length and breadth of the country then the necessity for discussing this question would not exist as it does to-day. But, Osteopathy being new and having but recently been discovered as a science of infinite value in the healing art, and opening as it does a most inviting field to all alike—to the good and worthy practitioner as well as to the unscrupulous adventurer, a few suggestions bearing upon the subject at this time would not be out of place.

Who should practice Osteopathy?

It will be understood, of course, that we refer more to the character of the person who undertakes this responsible duty than to qualifications.

Osteopathy should not be tampered with by an adventurer—whether man or woman—by persons who seek to make money by equivocal methods, that is, by employing the science simply to further his or her nefarious schemes. The adventurer is generally a bird of passage, ill-omened and unstable; taking up with anything, however pure or sacred it may be, provided they can use it to their own selfish advantage. They subject it to the working out of some unhallowed, individual problem, as a means to the end they desire, throwing honor to the winds.

It should not be practiced by those who fake their way; by those who adopt it only as a means for swindling. The world is full of fakes; they are found in all trades and professions, and in all occupations and avenues of life.

Whatever they can fraudulently manipulate for the purpose of getting something for nothing, and for the sole object of cheating their fellow-men, they unhesitatingly seize and push it so long as it yields any revenue to them. The fake entraps his victim and then filches the money from his pockets. He has no compunctions of conscience and will do almost anything that his courage will permit him to do to deceive his patient and to make the worse appear the better.

Osteopathy should not be followed by either man or woman who cuts the prices which have been established by the home institution as reasonable, that they may, by so doing, steal a patient from another worthy practicing osteopathic doctor.

This is one of the most sneaking, reprehensible acts that any man or woman can be guilty of, saying nothing of the man or woman who claims to be a reputable practitioner. No honest osteopath who has a proper regard for his own success would charge unreasonably for his services. But there are such people, I mean those who cut the price, to be found in all the professions; they are cheap men and women—cheap in every sense of the word. They are always on the lookout to find where a patient has gone or intends to go for treatment, and after they have spied out the fact they either make him a visit or lie in wait for him, and so insinuate themselves into his confidence that they worm the facts out of him, and then cut the price, not thinking, perhaps, that when they do this they are lowering themselves in the estimation of every well educated, fair-minded patient. These underhanded methods cannot be too severely condemned, because they are both undignified and illegitimate. They clearly show that the person guilty of practicing them is following the profession for only the money that is in it, and cares nothing for its

dignity or honor. Patients should be warned of these ser-
pent-like healers, for it is pelf they are after. It should always be borne in mind that cheap men and women do
cheap work.

Regular graduates in Osteopathy should not assist fakes,
for when they do so they are aiding in foisting upon the
public an imposter. No promise of reward, however great,
should induce a man who has obtained his diploma honestly to sacrifice principle and the good name of his profession by an attempt to boost another who is a fraud and
whose methods are those of a fake. When he thus lends
himself to sustain and build up the fortunes of a recognized swindler he becomes particeps criminis and should
be placed under the ban of all decent Osteopaths. He con-
fesses his own weakness and proclaims his own inability to
follow his profession in a legitimate way.

It should not be practiced by persons who are indolent
and hate work. The true Osteopath should be wide-awake
and full of energy. He should, when not engaged with
his patients, apply himself to the study of his cases, and
when he has no patient he should diligently devote himself to learning more of his profession. He should remember the scriptural passage, that the lazy and slothful
man, the drowsy man, shall be clothed with rags. This is
the substance of the quotation. It is as true to-day as it
was in the time of Solomon.

It should not be practiced by persons who tell more than
they know. The public is very apt to place a proper esti-
mate upon a man's ability and qualifications if he is a continuously loud and vociferous talker. A man may talk
and talk to the point; this is necessary in a professional
man; but when he talks both in and out of season he is
generally not rated very high as either a thinker or worker.
There are numbers of men in all professions whom the

ignorant and uninformed think are wiser than the men who stand at the head of them. The intelligent people, however, who know something themselves, will not only think less of the demonstrative man, but will unfortunately be inclined to think less of the profession he represents.

Those who start Osteopathic schools for the purpose of grinding out diplomas to sell. Such persons and the institutions they found are dangerous to any community, because they let loose upon the public a set of men and women who are complete ignoramuses, and who will do more to shake the public faith in Osteopathy than all other causes combined, because they are authorized by diploma to impose upon their patients. It should not be practiced by students who are not willing to battle for its principles at all times and under all circumstances; nor should it be practiced by the student who, when the good of Osteopathy is at stake, will lend his aid to help out the fakers and the opposition. If he does not try to save the reputation of the profession he follows he will not be very quick to save the smirching of his own garments.

So we might continue on through the catalogue of the unworthies who would bring reproach upon the fair name and fame of Osteopathy if they assumed the role of a doctor and practiced its teachings. But I have said enough to show the reader something of the scope that may be included when discussing a question of so much importance to the true osteopath.

We know that osteopathy will ultimately become like any other profession. Quacks and pretenders will multiply in proportion to the chances for making money out of it. But this condition may be warded off to a great extent, while the science is still in its infancy. if those who are true friends of Osteopathy will exert themselves along the lines indicated.

13

Upon the other hand, if those who are properly fitted by education and inclination to practice the new science will take it up as a life-work they are bound to succeed. They will always bear in mind that in proportion to their success osteopathy will be respected; not only respected, but it will grow as one of the potent factors in the art of healing. To make then a success of the profession the practitioner should endeavor to avoid everything that smacks of quackery. If he intentionally adopts any method of practicing or advertising that is not in harmony with the dignity and well-being of his profession, his efforts in behalf of the new science will militate against it. It is a lamentable fact that there are men in the practice of Osteopathy, as well as in medicine, who take up the mistaken idea that they would make money faster and more of it if they descend to the arts and methods of the quack. When they do this they are leaving the legitimate fields of labor and diverting the knowledge they have gained of their profession into improper channels. But they are altogether mistaken in their idea of making money. The charlatan who is recognized as such rarely ever becomes permanently located anywhere. As a general thing his patients are not those who are blessed with either money or culture to any special degree. They, like himself (or many of them at least), move about from one place to another.

The true Osteopath abides his time and he will not have to wait long either for his patients. He will soon become known to the community, and if he is a man of skill, as he must be, after having taken a thorough course in a school of Osteopathy, he soon begins to realize a constantly increasing practice. Take, for example, any physician who has lived for years at a single place and has been constantly in the practice. If he is fairly economical he has a fortune, either large or small, to show for his labor, while the shyster and pretender, although he may travel from post to pillar and constantly advertise himself, accumulates but little if anything over and above his ordinary expenses. Then if you adopt the new science of Osteopathy as a life-work, stand by it through thick and thin, and you will not only help yourself abundantly to this world's goods, but you will help your fellow man and Osteopathy.

SIX WEEKS UNDER THE RULE OF MEDICINE DISPENSERS.

One of the most graphically written and one of the strongest and most truthful interpretations of a man's thoughts, desires and feelings whilst lying for weeks on a bed of pain and anguish may be read in the subjoined article which was furnished by one of our recent patients:

I took ill on the 5th day of June, A. D. 1897. There was nothing momentous about it. No newspaper mentions. It was by no means a voluntary affair. It was no more feared than solicited. The demon bedfastness was unknown to me, and thus he had no terrors. Much like the fellow who was asked upon the cross-fire of a legal controversy if he paid the sheriff for fetching him ten miles to jail, replying: "No, sir; I did not send for him." The whole calamity came without any strong desire within me to try an experiment.

I say "calamity," yes, it is a calamity to be ill—bedridden—in any month of the year, or at any point in the ramifications of the isothermal lines. Sickness is no terror to the most active participant. He is busy guessing where and how to put in his time while confined to the narrow limits of the modern bed. The attendants are wondering, fearing, doubting, but the sick man has neither time to wonder, consciousness to fear, or vitality for a doubt. If there are qualms as to a hereafter let the man so possessed seek his Creator when in good sound bodily health as well as in the days of his youth.

As I said before, I did not get sick of a purpose. I even defied the wisdom of the family physician, and to tell the truth, I would have come off much pleased to have de-ceived him as to my ability to, as he termed it, "throw it off." Even his, "I said so," has a twang of comfort in it when he takes a perspective of the lateral dimensions of one's tongue. One is positively glad to have him come, sit down, look wise and confirm you in your suspicions that you are still alive.

I told my wife I would not be ill. I frightened the babies by the ghastliness of my assertions. Poor little "Bobsey," my six-year-old (scarcely till August 8th, 1897) wet his little cheeks with tears, whereof he could scarcely assign cause—save that "Popsey" was strangely unnatural, and took no heed of toys, and rolling the floor with him and "Sisser." Yes, and dear little two years old (not till August 17th, 1897), "Sisser," she could only exclaim in a wonderfully accentuated tone (a tone born of heaven, as I verily believe), "Papa," as she too saw that the strong (tut) and abiding (hush) one of the household was actually bending to break under the weight of something. Indeed, as I beheld the tears and heard the exclamations, I pictured Charles I. leaving his family, as the grim shadow of popular tyranny fell over his regal pathway. Ah, friend, there are none strong till they become weak. You can sort of sing in whispers: "Lead kindly light."

A collateral inquiry at the inception of a settled disorder of primal importance is, "What is the trouble?" There is enough disorder to disorganize a whole regiment, and yet to be able to locate and grapple the affection where it lives is a most momentous soliloquy.

Very soon the emaciated nurse, earning an asthmatic salary from off a Beneficiary Order (one of God's best institutions) becomes an object of envy. Where on earth is health and strength? Talk of doing things in the world! Why this man is a tower of strength, and soon one reposes in him with perfect composure. He not only understands the medicine, but he feels the pulse, takes the temperature, and takes the "long measure" of your lingual apparatus with very sober and intelligent composure. This man is the "second in command" and when the family physician has come, occasionally disputes with that sufficiently august portentiousness the efficacy and "soothing" powers of the "hop packs" over an unadulterated mustard plaster.

Think of it. Thermometer at ninety-six on top of the bed-post (no, at the foot-board—we have no posts on our beds, and I want to be exact in this matter), and hot "packs" 212 degrees Fahrenheit applied to the agonized abdomen, and a metallic fuser 12x12 across the "small of the back," meekly mentioned when first proposed as a "mustard plaster." To be flayed alive is merciful if done promptly.

Suicide is not a craze. It is not a species of madness. Life is an ugly mockery at best, and when the subject is well, and enjoying the allotted "three meals." Goethe, well, strong and prosperous at sixty, regrets being alive. At seventy-five, with the health and laudation of the world, wrote to Eckermann: "I have ever been esteemed one of Fortune's chiefest favorites; nor can I complain of the course my life has taken. Yet, truly, there has been nothing but toil and care, and in my seventy-fifth year, I may say that I have never had four weeks of genuine pleasure."

One who would thus complain, after being dined by dukes and received by kings, ought to have tried six weeks of servitude under the domain of the "science of medicine." Perhaps he did, but did he try hot weather, stringency in the money market, and an added aggravation and horror of one of the "Seven Plagues?"

Seriously (and sickness is a very serious business) the elements of a saint are prerequisite to a patient sickness. There are myriad things to be considered, as follows, in a medley of disorder: Business, social matters, engagements, medical aid, nurse, needs of the family, general finances, property, death and funeral expenses.

It will be borne in mind that the first day of my derangement was a day of unholy exclamations and surprises.

Possible calamity. Worse and worse and not better! Disease is a frightful monster! It begins its ravages with a stealth and insinuation that at every inquiry of the now starving system the answer is plainly negative. The whole diagnosis leaves the family and physician more and more in doubt and fear. The victim endeavors to ascertain the status of his malady from faces and voices. He cares least as to general results. True, he has much to lose, and nothing to gain. The fight is on. It is life or death. The decks are cleared for action. Injections are freely administered at various points of the sick anatomy. All the varied applications are brought into action, and on Sunday, June 6th, a miniature hospital is in full operation. The day is over and gone, with the patient in a comfortless daze, and the attendants still more so. Several days of intermittent fever and reeking perspirations. With each calamitous change I am convinced I am on the sure road to a change. The fever persuades the deranged mind that health and strength has surely come to wipe out malady. Presently the fever is displaced by reeking sweats. The parched tongue almost loses its office in the ocean of porous exudations. Then comes the collapse, and there is more than one fear of a sudden "taking off." The breath is labored, and the whole system is running like a neglected quartz crusher. The doctor arrives. He does not look grave. No, that would cost him a patient; but the patient he is practicing on sees through his disease the guessing sort of mal-administered look and feels more uncomfortable.

A change of medicine comes together with a complete

overthrow of the whole plan of action. Something must
be given again to relieve pain. After this the progress of
the unexplained distemper (save in polysyllabic inexpress-
ives) has the victim at its mercy. When the mouth is dry
the nurse can tell better than he. When an additional load
is to be given to the alimentary canal, he submits, and relapses with a death-telling groan. Say, have you been
through this? If so, is stands explained without multipli-
cation of descriptives.

There must come a climax. There have been many in
expectancy by the attendants. The physician has betimes
been dispatched for in hot haste. But nature soon burns
or wears out, like any other mechanism. Three weeks
have gone with their concomitant relapses and seeming
elevation; and now there is a chaotic state in which even
medical skill knows that the complaint or disease has
"spent itself." It was a down grade without brake or
steam to reverse. There is sure to be the final crash. It
has come. Forty-eight hours will determine whether it is
to be a long pull up the grade or the ostentation of a well
regulated funeral with all its moralizings and some distress. The engines of life are at work doing desperate
duty against the powers of destruction. The change is
slowly and certainly for the better. Now what are and
what have been my reflections? I have always been a busy
and energetic man, and time does not fly with its swift
wings without the mind has an avalanche of reflections and
variegated colorings.

Business, if you have any, is one of the primal agonies
of a season of distress of any kind. Not, will it suffer, but
how much? I never had the period of ill-natured recovery. I strung it along to the perpetual consternation of
family and nurses—save when real danger was pending,
then the victim went into that dreadful and dangerous

calm that augurs a short and easy road to what the German classic poet denominates Der Stille Land. While "business" is not everything, the American Goddess of Liberty should now represent every possible occupation known to the grasping, ambitious and cormorant American. Business is the God of the day and the night. It is the weariness of the one and the unrest of the other.

Presently the afflicted forgets it all. It was so with me. What of offices and stores in eternity? and sickness is an eternity terminal thoroughly commenced. One does not plan much of a paved heaven when broiling under disorders and heat. The "throne set in heaven" and the "sea of glass" are all as naught. It is a blank stare at anxious attendants, figures upon the ceiling, or darkness, dense, deep, awful, impenetrable, yet not frightful.

Once in a great while you will see a business acquaintance through the film of the unknowable, but he is not engaged in counting his ducats. He is in some kind of repose. The music you hear is in sweeter cadence—the words in rounded modulation—the earth, sea and sky in panorama as distinct as the illumination of Constantine on his illustrious mission. You feel much—but so much and so grievously that one cannot feel at all.

"The gay will laugh when thou art gone,
The solemn brood of care plod on,"

is one of the painful interims that steal into the strangely attuned ear of the afflicted. True, I had about me the dear babes and the fond wife, but sometimes "Holy Water" is not as refreshing as a draught of good old Burgundy. The one is a matter of faith and the other a matter of feeling—both written in the music of the "Psalm of Life." The world moves, and he who stands still will discover that Van Winkle has been asleep. While I lay prone at a mod-

erated full length no one in the stir of social life saw fit
to retard the wheels of social fellowship. I envied the animation and vigor that possessed the friend who could take
in the passing changes of our world of pleasure. I saw
opportunities with the forelock suggested by Samuel
Smiles pass and out of reach of the grasp of approach.
Woman, man, all whom I had seen in the garb of pleasure
flitted where "Fancy's magical pinions spread wide," and
I, alone, the solitary object of pity, and far from the reach
of envy or malice.

I do not care to drape the world in mourning. I never
like to see a hearse dragged over a gilded scene. I have
ever wished to see the smile, the laugh, the joy. What is
it to me to see? I can but think—the whole world sees,
but to me even hope is blinded.

Next come business engagements, or rather it all comes
together very similar to the old English Hotch-Pot Pud-
ding. A veritable conglomeration. It was all connected
—a sort of "rapid transit" vestibuled affair with "smoker"
and "baggage" included. I was on board, and yet dis-
patcher.

No one ever regretted so much a failure to keep an en-
gagement as he who learned thereafter of the much he
lost. Louis XIV said: "Promptness is the politeness of
Kings." It is one of the graces of an American, and especially a Denver "business man" to fail to keep an en-
gagement. Go, by chance, to the appointed place of meet-
ing and you are universally delighted with his prolonged
tardiness, enabling you to be gone before his arrival. What
one cannot do is exactly what he is in heat of want to do.
I fell to mourning the failure to meet engagements, for-
getting the many I failed, out of convenience or pretended
distemper, to promptly keep.

After advancing some days into the merits of my dis-

ease, the suspicion (and God knows suspicion is a bad
enough disorder of itself. Bad as ague, and quite as per-
tinacious in the most favored locality), for it was all of
that, stole over me that I might stand a fair and "fighting
chance" to live, provided the family physician was removed. I remembered my father's physician who solemnly
averred that his aged but robust patient must of necessity
die, being taken violently ill at the hight of the disorder.
He soon died leaving the patient short of medical affliction. It was noteworthy that all of the good man's patients
were (together with my father) able to attend his exequies,
despite the fact that he ailed but a day or two above a fort-
night. These dark promising, yet strange, reflections
passed between administrations as prescribed. The suggestions of friends were teeming with disfavor to the patient yet persistent family physician. He held to the same
medicines, and "repeat" got to be as "a sounding brass
and a tinkling cymbal." The time came for a secret
change. Again "man proposes and God disposes." The
elements contended for supremacy. The lightnings
flashed, the wind blew furiously, and the rain fell in tor-
rents. The able artists of the "knife" and "pills" failed
to make the tiresome six miles ride over well compounded
adobe roads, and "the evening and the morning were"
another day. It was cooler, and I could better withstand
the load of anaesthetics and calisthenics "prescribed." I
was, despite myself (including the family physician), de-
cidedly better. I continued to grow better till the good
physician again became, like Wellington at Waterloo, "the
hero of the hour."

It is said that good nursing is half the business in medicine. I go farther, it is two-thirds. The other fraction is
made up of the presence of the medical attendant without
his lotions and potions. Much of the patient's spare mo-

ments during lucid intervals is made up of curious, yet silent, observations upon the nurse. This unfortunate is impressed with the necessity of pleasing. He being much the only companion of the afflicted becomes his study and the "butt" of all the fun there is possible to elucidate from the critical dilemma. In the opinion of the untrained nurse the sick man is either better or worse according to his predilection. "I have strict orders not to allow any one to see him," was an invention the household respected, whether applied to the hungry collector who plied his occupation with a zeal equal to a better cause, or to the loquacious remedy-monger, who kindly has his own experience to vouch for the efficacy of the remedy. A nurse who "cannot tell a lie" may be a charming fable, but in actual practice or on "dress parade" is only filling his office by stipulation, and not with holy intent to bring the sick man out of his affliction.

The gravest question, however, that confronts one is not death, taxes, and the classified "botherations" of the sick room. "He that provideth not for his family is worse than an infidel." That is a short sermon, and like a true woman's postscript the appendix to be added to that assertion is the longer.

I have shed bitter tears, when on the couch of apparent death, I have seen my little slender boy strip for the night a very poor assortment of rags. There is a pity of anguish and an anguish of pity. Some, aye, much of our sorrows are imaginings; but that matters little if there is reality enough to make the soul desperately sick. The money short, the debts larger and accumulating, the small clothes worn threadbare, or worse, to the tender little skin, makes a theme worthy a philosopher. I once had the misfortune of throwing a whole carriage load of well fed and groomed men down a precipice near Pike's Peak. The remnants were gathered upon the road so foully missed, and another start made with renewed conveyances. I marveled at no word of complaint made from the passengers at their fellow-passenger's awkwardness. I so commented to them afterwards. One, a man of strange incongruities, yet possessing the grace of accomplished gentility, replied: "One would not be a philosopher who could not look with indifference upon an incident like that." Did you ever witness the pathetic child scenes delineated by Sol Smith Russell? Are you a father or mother? Have you ever shed tears at the privations of the street gamins? Think of your own dear little boy deprived. He need not be destitute—simply neglected. A father will rise from his deathly sickbed, he will grasp the furniture with a death grip while he draws about his now emaciated form his clothes and, with a haggard desperation in the absence of attendance, appear before the appalled family, lulled by the false aspect of sleep, a frightful specter, and in an anguish of feebleness exclaim: "My little ones must have a father's care."

Friend, have you been through some of that kind of life experimentation? I am aware that some strange creatures hate their own flesh and blood—that the mate of earlier summers has grown distasteful to their martyred morals and intelligence—that the kind of animation possessing them is a strangulated soul, a cold conscience, and a body alive to self and one self interest. But these are not natural. They are phenomenons.

I saw all this, felt it, acted it. But even the feeble baby hands stretched in supplication to a rash father importuning to have a care not to defy nature. I paid the penalty by a deeper prostration, and the soothing words of the one who above all others has stood the tests of time, and borne the burdens through the heat and heaviness of many years, were heard as in the far distance, resounding seemingly through some mystic cavern that echoed and re-echoed into silent whispers.

All this was a consciousness that the strong one of the house was stricken and become a "broken flax" and a reed swayed by slightest motion. But in our weakness we are strongest. I thought, as expressed by France's greatest fiction composer: "The supreme happiness of life is the conviction of being loved for yourself, or, more correctly, being loved in spite of yourself." It is strange to be happy on a bed of anguish. But there is in it some nearness to the good of earth and the better of an inexplicable elsewhere.

Never fear of a noble wife and loving children in adversity. They know the ebb and neap of the tide of affliction. There are more martyrs on earth than the writer of complex ethics ever dreamed; and when you make a martyr of a man, or even a small child, the devotedness and steadfastness of Moore, Latimer and Cranmer pale into insignificance. A strange philosopher was he who said: "God protect me from my friends—I will take care of mine enemies." He was a well man when he said this. Afterwards he suffered from gout. He was too prosperous to ever know what it was to be loved. There are martyrs who miss the crown but not the stake. These home martyrs are the makers of the real and ultimate good on earth. The man who goes down amid the waving of banners and blare of trumpets will be in the "dress circle" of true encomium, while that one who smoothes the pillow and closes the eyelids will have the choice of the parquette.

There is a vacancy of forlornness in an empty pocketbook. With nature all spread before one, and no helpless dependents, there is small need of worry; but wife, babes and domestic animals need, and their needs must be supplied quickly. An empty pocketbook is more painful than a "vacant chair." Thus I argued as the sum total of life insurance policies began to be weighed by the alchemy of sickness against the widowhood of a mother and the condition of the fatherless. With what grim anxiety is the footing of the profit-accumulating druggist scanned and the coin of the sunken pocketbook deliberated upon. There is a credit, but that has an embarrassing limitation. It is soon limited when distress is observed.

I have oft been so reckless as to asseverate, with malice aforethought, that if want confronted my family and I had no work for hands or brains to supply it, that I would —yes, I am sure, I have so expressed it with a firm and deliberate conviction—steal. Think—say to the belated: "Stand and deliver." After relieving him thus, being sure he could recuperate my broken fortune thus, I could look into the broad canopy of Heaven with the Image of God pictured before me and thank him for this bounty. This in nowise puts a pleasant countenance upon burglary, larceny or robbery. It simply follows the Bible doctrine that, "He who provideth not for his own household is worse than an infidel."

The great outside calamity to meet a man of affairs is money. His head, in the full enjoyment of health, is teeming with projects. His castles are erected high, and he tumbled over with them quite as often and much more heavily than the mentally athletic collegiate youth. There is much of brick, mortar and hard stone in the business man's castle of air, and when it goes over ofttimes wife, children and friends are scarred.

These escaping dollars, these fleeing dollars! When will they cease to flee beyond our needy reach? The sum total of a full payment is so managed to reach far, very far, so

that the mind is made sick with the trying to have all come
out even, or a little on the credit side.

The income of a man who lives on fees is sadly a vacuum
when he is locked in for repairs with "No admittance"
over the door. The evanescence of money is marked with
each personal observation. It comes to hand, and leaves
him with ease. Taxes are the cruel assessments upon personal industry. They are the drone producers and the
working bee's deprecators. Charge a man for being frugal
and provident and he soon, from that and other causes,
discovers that property is a menace to him. The vultures
that cling and hang about the prosperous man are thick
and untiring. Life is made a burden to him, for the reason that he has been made easy with a competence. He
had better be "passing rich" with "forty pounds a year."
Like Goldsmith's preacher. It is a dreadful chasm, this
of a happy mean between the extremes of riches and pov-
erty. Either extreme is a limitation of misery. The vor-
tex between is a seething caldron of half misery and total
oblivion. It is the disagreeableness and uncertainty of feeling of the somnambulist in partial recovery. It is the far
point from all realizations, and, yet, thus are the vast army
of the best industrial element. Most men and women
prefer a flash of unhealthy excitement to this humdrum of
routine. Thus suicide, dreadful murder and like senti-
mentalities variegated the once sedate lives of the diligent
and provident. More deaths, self-inflicted, occur from
some financial cause than from any one other cause. The
insane insatiable desire for gold, once known in youth never
leaves the adult. He clings to it more tenaciously than to
wife and children. The altar of property and possession
has more deformed infants and dead and emaciated wives
than the fires of the heathen gods. Could the image of
Vishnu see the seething and mangled mass of broken flesh

strewn behind the army of gold seekers he would stop his blood-stained wheels in sheer discouragement. Verily I have thought thus, upon my bed of disease, that the love of this world's gains is the root of all human evil.

In the still of night when naught disturbs but the restlessness of the tired nurse, and the sigh or groan of the sick; when the walls present something of the unearthly aspect of strangeness; when the insects without seem to chirp the very beatings of a lapsing heart; when all concentrates its burden of hopeless abandon, and the dread darkness seems to settle down like a pall, to say: "Sleep, this is death." Then it was that I felt—and forever to feel —that there was only despair in this world of possession-loving. The kind loan of to-day is the oppressive debt of to-morrow.

On a bed of sickness what, friend, have you ever thought of death? Did you ever go to the cemetery and select for yourself a cold resting place? You have been able to do that for others, but not for yourself, save in the flush of health or in the melancholia of some oppressive ailment akin to sickness. There is an appalling indifference upon this matter of death. Queen Victoria held her multi-million pounds jubilee while 20,000,000 of her subjects were starving by a long drawn out torture in India. Murat, Robespierre and Danton at the French guillotine, Warren Hastings over the primitive miseries of England's unfortunate protege, and the European Nero rasping his fiddle while Rome went up in smoke, are each and all feeble, gentle and gracious examples of mercy and charity as compared with the gaudy mother queen jubilating while death stalked everywhere among her dusky subjects, who were better in savagery than in the doted bastard Christian affiliations.

If God has any use for men he knows best when and where to use them, and there is small cause for alarm if

14

he does not respond to each eager exclamation: "Here I am, send me." If death is to come the least conscious of the end is generally the dying. Some one has said: "I am not afraid of death, but I am afraid of dying." This dying is a simple problem. You are in it with no worry for pasts or futures, the eyes close and there is no whispered prayer. When the invalid prays, look for a speedy recovery. Death is of itself an ample employment for every faculty. The dying have no time to discuss eternity. He is busy on this side trying to keep up a pulse and respiration.

I had more trouble collecting atmosphere enough to formulate a good honest breath than I did in preparing any scheme of "What, after death?" If you have any idea of Heaven and Hell, friend, get at that and have it settled before you are physically deranged or mentally unbalanced. It is as impossible for one to get ready for Heaven, in my humble opinion, in the few disturbed hours of sickness before death as it is for the "Foolish Virgins" to make oil and prepare their lamps at the last moment. I have ever been a firm believer in Bible doctrine as it reads, not as the heavy divine expatiates upon it, but as God made it read in its simplicity for men. But I have never been brought to the fatal edge of death with any notion that after getting on the cross full forgiveness would likely follow. Besides religion is too difficult a problem for the one who is already weighed with disease. It takes a well man to digest it. Egg-nog was difficult enough for me, and I allowed the differences of belief to remain between Ingersoll and Dr. Talmage. As I am not yet strong, I shall leave much of that indigestible matter for those who will persist in overloading themselves with such forbidden fruits.

At every sign of recovery I found I must read. Not so

much to prepare for usefulness here but a diversion, and at the same time a vague notion of trying to associate with noble thoughts. I read Rasselas. It taught what I felt I knew: The uselessness of being alive, save in so far as life furnishes to the liver some strange experiences. The book is poorly constructed—as a fiction, useless. But the sentences are substantial, and they are good for those who are discontented. It is a homeopathic medicine. It fights fire with more fire.

William Dean Howells, in his Impressions and Experiences, writing upon "An East Side Ramble" in the poverty-stricken quarters of New York City, says: "I found them unusually cheerful. * * * * And they had so much courage as enabled them to keep themselves noticeably clean in an environment where I am afraid their betters would scarcely have had heart to wash their faces and comb their hair."

I am no admirer of Dean Howells. He may have enjoyed the pleasure of washing his face, but he never combed his hair, that is certain. But it teaches that, whether the promise of the earth is to possess lords and serfs or not, the masters are no happier than the slaves. The one can ill envy the other his lot. The happy quiet of a bed of sickness is the nearest to a paragon of bliss a man can have here on earth after all. Not that in the "throes of mortal pain" he revels in ecstacy! Not at all. The utter abandon of all to an inevitable makes all look easier. He is not disturbed by politics, bad theology or that death to all earthly pleasure—business. He is for once monarch and master. The house moves at his will, and even the small ones soon learn to revere the cot upon which their oft ridden hobby horse is, perchance, dying.

Did it ever occur to you, friend, that in sickness, not a moderate green-apple or over-early-melon-gripe-sickness, but real death-dealing bedfastness, that more time is devoted thinking of funeral expenses than of the much exaggerated "over there?" Not that dying men are penurious. Suicides think more of the dispositions of their remains than they do of the confronting perdition. There is no meaner debt than the price of a coffin and shroud, both useless, and should not be worn except in the most extreme cases. Think of a collector coming into your place of business with such a bill. It makes one shudder and become pale.

We know that sorrow for the dead is the only sorrow that will not be assuaged. We feel the ease of dissolution easier than the struggle for life and bread. It is a hero who dares to live, and all of us are cowards enough to die. The inquisition has no terrors, the storm no alarm, and the executioner no fears. It but wants the calmness of fate to say: "Thy will be done."

There is an unexplained and unrecognized felony in the fault of thinking too hard and laboring against the flood of human destiny. The happiest moment is that in which we are compelled to think. Pope, the afflicted, was full able to feel and thus to sing:

> "Our proper bliss depends on what we blame.
> Know thy own point; this kind, this dull degree
> Of blindness, weakness, Heaven bestows on thee.
> Submit, in this or any other sphere,
> Secure to be as bless'd as thou canst bear;
> Safe in the hand of one disposing power,
> Or in the natal or mortal hour.
> All nature is but art, unknown to thee;
> All chance, direction which thou canst not see;
> All discord, harmony not understood;
> All partial evil, universal good;
> And, spite of pride, in erring reason's spite,
> One truth is clear, whatever is, is right."

In this affliction of the soul and the senses, what was to do? Business could not be resumed, after months of bedfastness, and a convalescing that was even worse than the inertia of total disability. The office must be closed, and, worse than crape on the door, for a man who is in the fight for life and bread, for self and family, is physical decrepitude, with its accompaniment, mental inaptitude.

While slowly plodding towards a home, desolated by the failure of the one from whom all the confidence should emanate, and with that settled darkness overshadowing all, the chance (are those things chance?) of crossing the path of one of Denver's most capable and practical preachers drew out a few words touching upon the amenities of life and the platitudes of illness. "Why not go and see my doctor?" was a common-placed inquiry. "Who is he?" was a natural inquiry. "W. L. Harlan, and his offices are in the Kittredge Building."

Many people have blessed Myron W. Reed for his plain naturals of life; but, without being an advance agent, for any new speculation and simply to do good to an afflicted friend, he dropped this response with more than ordinary emphasis. He passed on to sermonize, and I to a new determination. I went to see Harlan. He, to me, appeared plain, unassuming, ordinary. The first treatment gave me inspiration to open my office and resume business. From that time on life has seemed easier and business brighter.

DESCRIPTIVE.

As the Rocky Mountain Infirmary of Osteopathy is located in Denver, it may be interesting to persons who have never visited the city and who contemplate being treated at this institution, to know something of the location of the city, its climate, its environments, its places of amusement, its hotels, its street car and railroad facilities, schools and churches, club life, fine arts, water supply. cost of living, etc. We therefore publish the following descriptive article:

DENVER.

The Western metropolis and state capital of Colorado is in point of physical diversity of climate in its surroundings and proximity, as well as its diversity of occupation, the most wonderful city on the western hemisphere. As was said by a competent writer: "We must beg our stranger friends, who come looking for sights, to remember that only a generation ago there was nothing where Denver stands but cactus, buffalo grass, wild animals and the red man. The foot of civilization had not pressed the arid soil, nor had the magic hand of human genius been laid on a single thing, living or inert. That he should find, 600 miles west of the Missouri river, beyond the 'Great American Desert,' a city of 150,000 inhabitants—a city with 150 miles of rapid transit—a city of such beautiful homes, and public and business buildings as meet his gaze at every turn, is sufficient to amaze the thoughtful visitor as he comes among us. The work of the magician who, by the touch of his wand, can cause flowers to bloom from an empty glass in his hand, before your face and eyes, pales before that finer and more magical touch of civilization which has caused this city to spring up out of these arid plains, with so many miles of beautiful streets, shaded by grateful foliage, lining our sidewalks on every hand; with all these churches and school houses. Denver, in all that it is and promises to be, is more wonderful than mountain peak or canyon—they are natural, it preternatural. Denver is situated in the valley of the South Platte, twelve miles from the foothills of the Rocky Mountains. The site of the city is neither level nor hilly, but pleasantly diversified by a rolling and undulating surface. The distance from the mountains is sufficient to give a fine view of the range and a landscape for magnificence without parallel. The range is seen to best advantage from the end of Eleventh Avenue. Bayard Taylor said, when visiting Denver in 1866, that from this hill could be had one of the finest mountain views in the world. The range is visible for 200 miles. Pike's Peak, 70 miles to the south, when the air is clear, seems hardly half so far, and Long's Peak, nearly as far to the northwest, appears at times but a brief walk distant."

HEALTH RESORT AND CLIMATE.

The climate of Colorado is a mecca of health and sunshine which the despairing invalid ever seeks. The germ of disease finds no place in the midst of its mountains and plains. In the sultry and hot days of summer in the middle Atlantic and Mississippi Valley states its inhabitants are driven by the thousands to Denver and its vicinity, where, "freighted with the fragrance of mountain skies, wafted by pure, crisp, energizing breezes, the ozone of Colorado invites hope, happiness and health. From her soft blue skies an ocean of sunshine falls upon the fair face of Colorado. Nature could no further go to make a matchless climate. Of all climates known to be suitable for the checking of disease and capable of effecting a cure, that of Colorado stands unsurpassed. A rarefied air produces a peculiarly beneficial effect upon the diseased lung tissues; while dryness and sunshine keep the invalid buoyant and constantly toned up and warm, and fortifies him to withstand despondency and disease." Denver's proximity to the mountains on the west, its gently sloping plains to the eastward, its mean latitude of extremes of temperature, especially fit it as the center of the health giving region. The sterling and energetic people who have developed Denver would have done so in any event, owing to its manifold resources, but it is evident that its magnificent climate has acted as a magnet in attracting to Denver masters of all crafts. Its park area is more than that of the historic city of Paris. City Park, the popular common of the city, covers an area of 320 acres. It is adorned by shade trees, shrubbery, flowers, lakes and famous statuary, and is but twelve minutes' ride from the center of the city. In the summer months, daily concerts are given at the expense of the city, by a fine brass and reed band, and at certain hours of the day its capacity is well taxed. Lincoln Park on the south and west part of the city ranks next in attractiveness. Among the others are Dunham Park on the east, Chaffee Park on the north, and Congress Park, the Capitol Hill high point, containing 160 acres. Besides these, Park Avenue intersects ten streets diagonally, and at each intersection forms triangular plots which have been improved by shrubbery, flower beds and "seats beneath the shade," making twenty-five little parks. In the suburbs of the city, but a twenty minutes' ride, is the Elitch's Garden, one of the most delightful resorts in the West, and it

THE AMERICAN SCHOOL OF OSTEOPATHY.

...*DIPLOMA*...

Know all Men by these Presents, that William L. Harlan having attended a full course of Lectures on and demonstrations of Osteopathy, and having after due Examination been found fully qualified to practice the Art in all its branches, is hereby conferred by me with the title DIPLOMATE IN OSTEOPATHY.

Dated and given at Kirksville, Adair County, Mo., this the 2nd day of _____ Eighteen Hundred and Ninety _____

A. T. Still
PRESIDENT AMERICAN SCHOOL OF OSTEOPATHY

SECRETARY

has been aptly and deservedly denominated "The Pride of Denver." It is the refined breathing spot and public amusement place of the city. Here is to be found the home-like, restful park and flower garden, just the place of all places where one is benefited by a few hours' vacation, and finds restful music, fresh air and geological specimens precisely to his or her liking. Here is a refined and high-class theater, with a seating capacity of 3,000, where two performances per day are given to patrons of the Garden. The Zoölogical Garden is filled with animals from every clime and continent, and affords a full share of amusement to the passer-by.

Within a few hours' ride from Denver can be visited on excursions, at low rates of fare, the famous "Loop" up Clear Creek Canyon, Colorado Springs, the clean, refined and classic modern resort of the West, Manitou, designated the Saratoga of the West, with its outlying scenes and attractions, the principal ones of which are the Garden of the Gods and Ute Pass, up the canyon.

The dryness of the air of this great mid-continental tableland, and the consequent rapidity of evaporation, must be kept in mind in considering Colorado's temperatures, would one gain an accurate understanding of the climate as one feels it. The average July temperature of Denver is 72.1 degrees. The Denver summer corresponds as to the feelings of those who pass through it, to that of Manitoba, of the Thousand Islands, of the Adirondacks, or of the White Mountains (Capt. Glassford). The summer heat is seemingly occasionally intense; but it is really little felt, causes very little inconvenience and never any suffering. In the hottest of summer weather it is but a step from the heat of the sunshine into the shade, which is always cool. Sunstroke is here unknown. This coolness in the shade of Colorado, due to the very rapid dissipation of heat by reason of the rarity of the air, is something often spoken of but not easily impressed sufficiently on those not familiar with it. It makes it possible for one to live with great comfort even during the summer when the general temperature, as shown by a thermometer exposed to the direct rays of the sun, would seem to be almost unbearable. The summer nights are always cool and one can always sleep under quite a heavy comfort or blanket.

VIEW OF MOUNTAINS FROM DENVER.

From the dome of any of Denver's magnificent business blocks may be seen some of the grandest mountain peaks of the Rockies; snow capped Pike's Peak, 14,147 feet above the sea level, seems but a few miles distant; from Denver may be seen the mountain upon whose side is located the deepest gold mine in the United States, the California, while in the opposite direction one can easily see the hills that surround one of the world's richest mines, the Portland. Every day the citizens of this great city see at least seven mountains from whose rock crowned tips the snow never disappears. No city in the world commands the number and variety of views of noted and historic mountains as does Denver. At all times the mighty Rockies loom up as monuments to that unseen power which created a perfect clime.

The mountain ranges, spurs, divides, plateaus and mesas make up the vast network of Colorado's mountains. The eastern plains of the foothills are 5,000 feet above the sea level, and out of this immense plateau rises a vast tangle of mountain peaks and chains and table lands. From Denver can be seen the wide plains, the pleasant fertile valley, the rounded hill top, the abrupt peak rising to 14,000 feet, the steep cliff, the narrow gorge, the butte or monument of solid rock and the close shut in valley, all of which, by a preponderance of nature's prolific endowment, are found by the development at the hand of man, to contain the riches of gold. It can be truthfully said of the Rockies, that they outrival the Alps of Switzerland.

PLACES OF ENTERTAINMENT AND AMUSEMENT.

During six months of the year, Denver is pre-eminently a city of out-door amusements and pleasure seeking; but by virtue of the many parks, resorts and outlying places of interest and amusement, its large population is so perfectly absorbed and conveniently cared for as though it were an every-day occurrence. Muddy streets in Denver are a rare occurrence, and it has been estimated by cyclers the past year that there is good wheeling for at least 300 days in the year, and the bicycle on our finely paved streets and gravel roads is constantly in evidence.

There are four theaters giving regular entertainments in the city during the theatrical season. Foremost among these are the Tabor Grand Opera House and the Broadway Theatre. The former of these, built in 1880 at a cost of $1,250,000, stands as a monument of beauty of design and architecture of Denver's early ambition and greatness. It has a seating capacity of over 2,000 people, and the highest class of attractions are the rule. The Broadway, built in 1890, in connection with the Hotel Metropole, at a cost of $520,000, is an imposing structure of eight stories, and has a seating capacity of 2,500 people. During the past year a stock company played in the Broadway to crowded houses.

WATER SUPPLY.

It is seldom that people consider how active a part a plentiful supply of pure water performs in the upbuilding of a great city. The original water system of Denver was started in 1871. Various improvements were made from that time until 1889, culminating in the latter year in the splendid system now in use. Denver water now comes from the mouth of the Platte Canyon where there is practically an unlimited quantity of the most excellent water. This is conveyed to Denver in two water conduits, one of which is twenty-one miles long, and enters the city on the north side through the city of Highlands, and the other, which is twenty-three miles long, enters the city on the east side through the University Park and Capitol Hill. Every effort is taken to delivering the water to consumers in a pure condition, and on that account great care is taken that every gallon is filtered either by natural or artificial means. The water is correlated by means of ditches and flumes running high up Platte Canyon, where the snow water is pure as crystal, and accumulated in reservoirs, one of which covers an area of over 400 acres and 35 feet deep, and from the reservoirs it is thoroughly filtered through the filtration system of C. P. Allen, engineer, and thence carried in the two large conduits mentioned to the city by gravity pressure, being thus free from contamination and pollution of all kinds, and furnishing adequate pressure for all demands.

HOTELS.

By virtue of Denver's geographical situation, as well as its proximity to the manifold physical attractions in the surrounding mountains, it has gradually grown to the very

first rank of national convention cities of these United
States, and there is not a week passes but what a national
convention of some sort, either religious, scientific or artistic, is held in this city, and this feature, as well as a very
large summer tourist travel has made an exceedingly large
hotel demand necessary to keep pace with it. There is not
anywhere an hotel more simple and artistic in appearance
than the world-famed Brown Palace. It was opened in
1892; is built of brown sandstone with carved decorations;
has grand rotunda finished in Mexican onyx; is ten stories
high, has more than 600 rooms, and cost $2,000,000. The
other leading hotels are: The Windsor, the Albany, the
St. James, the Oxford, the Imperiale, the Metropole and
the Columbia. The American House, owned and conducted by Charles Smith, is the veteran hostelry of Denver. An influx of 100,000 people into the Capitol City of
the Plains from the outside world are as comfortably absorbed and cared for by its hotels, boarding houses and
rooming apartments, and dined at its many restaurants
as though it were an every-day affair.

STREET CAR FACILITIES.

Denver, covering as it does a vast area of territory, with
the many home-like incidents by way of lawns, roomy lots
and breathing spots, makes it a very large city in proportion to the number of its population. The city, from its
business center, extends for miles in every direction and
its suburbs are the more sparsely populated. To keep
pace with the prime necessity of transportation, western
enterprise has not been wanting; for gravitating at a common center, extending from 17th to 15th streets, is a
street car system not excelled anywhere in the land. These
superb lines of street railway radiate in every direction,

running to the utmost limits of the city, as well as to the suburban towns miles away. There are now over 150 miles of street railway in Denver, and there is not a locality but what has easy access to this modern method of rapid transit. The lines are for the most part electric motive power of the most approved methods.

Every night of the week (Sundays excepted) can be seen on the electric lines dozens of tourist and pleasure cars, illuminated by variegated colored lights, excursion parties made up by colleges, schools, churches or clubs, who are given a 25-mile ride for twenty-five cents to the different parks and attractions about the city limits.

RAILROAD FACILITIES.

Denver is a railroad center. Fifteen railroads operate in Colorado, and all have their main offices in Denver and radiate from this point. There are 8,000 railroad employes residing in the city of Denver. The total trackage in the state is 4,780 miles. A few hours' ride from Denver in a Pullman palace car over the Denver & Rio Grande Railroad would bring one to the most picturesque mountain scenery in the world, and enable one to go to the highest mountain passes in the Rockies, and through the deepest and wildest gorges; for this road has become noted the world over for marvelousness in construction, wonders in engineering, grandeur of scenic attraction and universal excellence of equipment. The first railroad was completed into Denver in 1870, and since that time all the great lines west of Chicago have either lines of their own to Denver, or connections. Fourteen through trains reach Denver daily from Chicago and eight from the Pacific coast. The Denver union depot is a granite structure covering two entire blocks; the building and ground having cost a half million dollars. It is a marvel of beauty and architecture. and is the common center of the passenger traffic of all the roads running into Denver.

SCHOOLS.

For educational advantages, Denver is ‾unexcelled by any other city in the United States. In the year 1896, Denver spent $804,385 to maintain her public schools. Among her institutions of learning are: East Denver High School, which covers an entire block, and cost $360,-000. In the public library are thirty-five thousand volumes. The North Denver High School, and also that of West Denver are most admirable and costly buildings, with all modern appliances and improvements. The Manual Training High School was opened in 1893, and cost $100,000, and is one of the most complete in the country. The institutions for higher education are: Wolfe Hall (Episcopalian) for young ladies; Denver University (University Park) has the well-known Chamberlin Observatory, which cost $100,000, is a Methodist institution; Jarvis Hall (Episcopal), located at Mont Clair, is a boys' Academy; Jesuit College on the north, and Sisters of Loretto Academy on the south; Westminster University (Presbyterian) in North Denver; the main building was erected in 1891 and cost $200,000. There are many other minor institutions and business and commercial colleges. The public school enrollment of Denver is, ward schools, 25,771; high schools, 1,683; Manual Training High School, 250; kindergartens, 3,500, making a total of over 31,000. No city in the United States has better facilities in this line than the capital city of Colorado.

In this rapid development of Colorado, nothing is more noteworthy than the growth of its educational interests; not only elementary and secondary education, but higher education has shared in this growth. The last German year book of the Educational Worlds rank the University of Colorado with the first eleven American universities, and the first four state universities. Eastern students seeking a change of climate find in our schools privileges equal to those at home.

CHURCHES.

Denver has some of the finest church buildings and places of worship in the West. There are 121 church organizations in the city and 90 church buildings, outside of the various missions situated in the suburban portions of the city. No creed known to man but that the believer can find a place of worship in this city. The most notable church edifices are the First Baptist, Central Presbyterian, Trinity Methodist Episcopal which cost $280,000, St. John's Cathedral, St. Mark's Episcopal, Temple Emanuel (Jewish), Unity, the new St. Elizabeth and the Christian. The Christian has a number of large churches in the city, the principal one of which is on Broadway, between 16th and 17th streets. The value of the church property situate within the city limits is $5,200,000; the total membership, 47,500; about 150 ministers are engaged at salaries of over $1,000 per year. In our large churches are found some of the most noted and eloquent expounders of the doctrine taught by christianity.

CULTURE OF THE FINE ARTS.

On account of the diversity of landscape, the immense mountain peaks interspersed with undulating hills, valleys, through which run the rippling streams and dancing brooks through the rocky gorges, with all shades of colors and scenic effects by way of shrubbery, makes Denver and its immediate vicinity an especially desirable field for the exercise of the artistic talent in all its phases. In this city are many of the most reputable and talented artists from the eastern centers, who have migrated to this fitful field of nature to teach and practice the art of reducing nature's sublimity and magnanimity to canvas and to teach the same to students. This fact is now so universally recognized that all through the eastern and middle states Colorado scenes are reproduced in the public schools and colleges for the benefit of students.

Not only is the painter on canvas and photographer in this region strongly in evidence, but dozens of studios and academies are conducted for the instruction of oratory and elocution in all its departments.

CLUB LIFE.

An incident to the growth of the Western Metropolis in all its various channels and avenues, that of club life and club building occupies a conspicuous place. The clubs are equal to that of any city. The Denver Club, 17th and Glenarm, is a massive red and lava-stone structure, built about twelve years ago, at a cost of $250,000. The Denver Athletic Club, 14th and Glenarm, is one of the finest in the country; building and equipment cost $225,000; membership, 1,000. University Club, 17th and Sherman Avenues, has a membership of 200; the building recently finished is of white and gray tile brick, colonial style of architecture and cost $20,000. The Progress Club (Jewish) has an elegant club house at 2047 Lincoln Avenue; the material used is undressed lava-stone. The Woman's Club of Denver is a most progressive organization, has more than 600 members, and is to build a modern club house. The Denver Wheel Club, 550 members, is erecting a new building.

15

COST OF LIVING.

To the man of means and to the poor man Denver offers every variety and extreme of pleasure and comfort, being circumscribed only by a person's desire and ability, concomitant with one's means. On account of our various altitudes, Colorado produces not only the lucious semi-tropical fruits found in California, but also the staple cereals and root crops found in the central and northern states of the Union. The climate, always pleasant and sunshiny, and free from destructive storms, is exceedingly favorable for producing all grains, grasses, fruit and vegetables that grow in the temperate zone. Denver being the trade center of the state, here are handled all of the products raised, and consequently with the result that eatables in the form of farm products and fruits are cheaper than in any other city west of the Mississippi river. Denver, for a number of years past, has grown to be one of the convention cities of this country, and on that account is able to handle with convenience and ease more people comfortably than any other city in the land. Owing to the mildness of our winters, to clothe one's self is not an expensive task, as a medium woolen suit is suitable for either winter or summer wear. Clean, healthful living apartments, with board, can be had at prices ranging from $3.00 per week to $40.00 per week, according to the degree of luxury, convenience and comfort desired by the sojourner.

Osteopathy has friends among the best and most distinguished people of our country, as the following letter will show. It was written by Mrs. Julia B. Foraker, wife of Senator Foraker, of Ohio, to Dr. A. T. Still, of Kirksville, Mo.:

COMMUNICATIONS.

The gratitude of some of our patients may be seen in the following communications. They were written with the hope that others similarly afflicted might have the benefit of their experience with Osteopathy, and are published herein only at the request of the patients:

"I have been much interested in the science of Osteopathy as reduced to practice by W. L. Harlan and assistants at rooms 401, 402, 403, 404, 405 Kittredge Building, Denver, Colo. I remember the line in the familiar hymn of Dr. Watts—'Strange that a harp with a thousand strings should keep in tune so long.' Dr. Watts must have known by experience and observation that this mortal machine does not keep in tune. Man, considered as a machine, is more intricate and delicate than any harp. He is exposed to extreme heat and extreme cold—he is under the weather. No violin has such an experience of dry and wet, noon and midnight. It is not strange that he gets more or less out of order. The wires of the best telephone system become crossed and tangled. I suppose from what I see, bones, arteries, veins and nerves may easily get in one another's way: They must be placed where they belong. This may be done by the trained hand of a man who knows where they belong. It is simply this, the instrument must be tuned. I observe on the street the people, there is something the matter with many of them. I see that one foot drags, that one shoulder is higher than the other, that the clothing along the backbone does not hang plumb. The machine needs manipulation, a setting to rights. I do not suppose that any drug will have any influence on bones out of place. I suppose that the lungs and heart will work better if they have plenty of room.

"I have been interested in Mr. Harlan's work, and anyone will be who sees him do it. I say cheerfully that I have been benefited by his work on me. He is a gentleman who knows his business. And my friends who are unequal, one-sided and out of gear can by him be put in right balance and working order.

"Osteopathy is a new thing, but so was the discovery of America. It is the end of the century and the time for new things."

"Monument, Colo., Oct. 26, 1896.
"Dr. W. L. Harlan, Denver, Colo.:

"Dear Sir: Last August my daughter, Mary Evelyn, was afflicted with facial paralysis, which came on very suddenly. She had no control whatever of the right side of her face. I consulted three of the best physicians in Denver and they advised me to take her to sea level at once.

"Through the advice of a friend I took her to you on the second of September and made arrangements for a month's treatment, and to my surprise, after a few treatments, she was entirely cured."

"Denver, Colo., March 10th, 1896.
"To W. L. Harlan:

"I have been under your treatment for a period of four weeks, and am now feeling quite a new man. At the time I was examined by you I was suffering from severe pains in the head; also from a torpid liver and stomach troubles, appetite bad, with sleepless nights. My kidneys gave me a great deal of trouble, urine contained quite an amount of uric acid. I had, also, pains in the back and hip due to an accident. You reduced my superfluous flesh seven pounds in four days. My appetite began to improve and my sleep became sound, all due to the new treatment of disease."

"Denver, Colo., Aug. 16th, 1896.
"My Dear Mr. Harlan: You ask me for my opinion of Osteopathy. You, who are the able exponent and practitioner of the science, leaving nothing to be said, but having been directly benefited by the treatment, my recommendation may carry weight to the afflicted. I have suffered tortures for years with asthma and bronchitis, which a Chicago specialist pronounced consumption. I had the

most violent paroxysms of coughing, which lasted for hours, and was very much discouraged when I came to you. I have taken four treatments. After the first treatment I did not cough once for nineteen hours. The asthmatic spasms have disappeared. I cough less and less. I eat as heartily as I ever did, sleep soundly all night, and, best of all, feel the buoyancy of spirits which betokens returning health and vigor and improved circulation. Even the skeptical who see how I am improving are being speedily converted to Osteopathy."

"Denver, Colo., Dec. 22nd, 1896.

"Three years ago I strained my back, and in a few months it developed into sciatic rheumatism. I took all kinds of medicines for the first few months, but it did not do me any good. Last August I was taken very bad. I was drawn out of shape and had to be wheeled around in a wheel chair. I tried Hot Springs and everything I could think of, but to no purpose. I finally heard of Osteopathy, and after looking it up well I began to take treatments and am improving right along. My spine is about straight and the muscles are relaxing, so that I expect soon to be well, and can cheerfully speak a good word for Osteopathy. I think it is a wonderful treatment."

"Denver, Colo., Nov. 23rd, 1896.

"Dear Sir: I have been afflicted with rheumatism for the past twenty years, and have doctored with quite a number of physicians. One would give me one thing and another something else as the cause of my trouble, and would doctor me for it, but I never received any permanent relief. I tried all kinds of patent medicines that I heard of, that were recommended, but all with the same result. At last I was persuaded to try your treatment for three months. When I began I was not able to raise my left leg off the floor while sitting, and was drawn out of shape very much and suffered great pain all through my body. I also had a very severe cough. But now after trying your treatment for three months, I can say that my rheumatism is almost entirely gone, my limbs are entirely free from pain, and I have the use of them as well as a man at the age of sixty could expect. My cough has improved. I can heartily recommend your treatment to any one who is afflicted as I have been. I believe it will do more good than all the drugs one can take."

"Denver, Colo., Aug. 26th, 1896.
"Dr. W. L. Harlan, City:

"My Dear Sir: I will state that for the past twenty-three years I have suffered greatly from rheumatism, much of the time being totally unable to attend to business, and having on several occasions been confined to my bed for months, and once pronounced incurable and given up by my physician. I have been treated by more than a dozen physicians, none of whom could give permanent relief. For several years past I have been unable to turn over in bed at night without severe pain, which has broken my rest so constantly that life was a burden. I began your treatment three months since, when I was hardly able to walk, and could not use my arms to dress myself. I have taken no medicine since, and I am in better health to-day than I have been for years. I have no pain night or day, and I sleep as soundly and comfortably as a child. I walk on an average four to five miles every day without pain or fatigue, and I have just been in the mountains examining shafts and underground work for hours at a time, and am apparently sound. I need not add that I am enthusiastic over Osteopathy, and my advice to rheumatics is to take the treatment."

"Denver, Colo., August 26th, 1896.
"Dr. W. L. Harlan, City:

"Dear Sir: Upon the suggestion of a friend who had experienced beneficial results from treatment administered by you, I was induced to give your art of healing a trial, after having tried various remedies prescribed by physicians for dyspepsia, with which I have suffered for the past four or five years. I have had five or six treatments at your hands (hardly a sufficient number to thoroughly test your system of treatment), and while I am unable to say that I am fully cured of my ailment, I can say that I have derived material and noticeable benefit, and expect, with a few more treatments, to be restored to my normal condition."

"Denver, Colo., January 8th, 1897.
"Dr. W. L. Harlan:
"I commenced treatment with you in September for muscular contraction in my knees, lumbago and cold feet. After two months, I find I am greatly benefited, and can very cheerfully recommend your treatment as highly beneficial."

"Denver, Colo., January 7th, 1897.
"Dr. W. L. Harlan:
"Dear Sir: To my mind, the Osteopathic treatment is the most reasonable and common sense method of curing diseases, and I have derived much benefit from it in a short time.   Srs. of Charity."

"Denver, Colo, January —, 1897.
"Dr. W. L. Harlan:
"For two years I have suffered from a pain in my hip which the M. D.'s said was rheumatism. A friend told me to see you. I did so. You examined me and found the trouble to be in the spine, and after a month's treatment I was entirely cured. I shall be glad to see or talk with anyone who is afflicted as I was."

"Denver, Colo., January —, 1897.
"To Whom it May Concern:
"This is to certify that I am acquainted with Dr. W. L. Harlan, having been treated by him at his infirmary. I know him to be a good anatomist, well posted in Osteopathy, a skillful operator and a man worthy of your confidence.   State Evangelist."

"Denver, Colo., January 22, 1897.
"Dr. W. L. Harlan, Osteopathic Infirmary:
"Dear Sir: Early in March, 1896, I fell from a carriage, sustaining injuries in my right hip and knee, which disabled me so I was unable to walk. A physician was consulted, who placed the member in a splint, where it re-

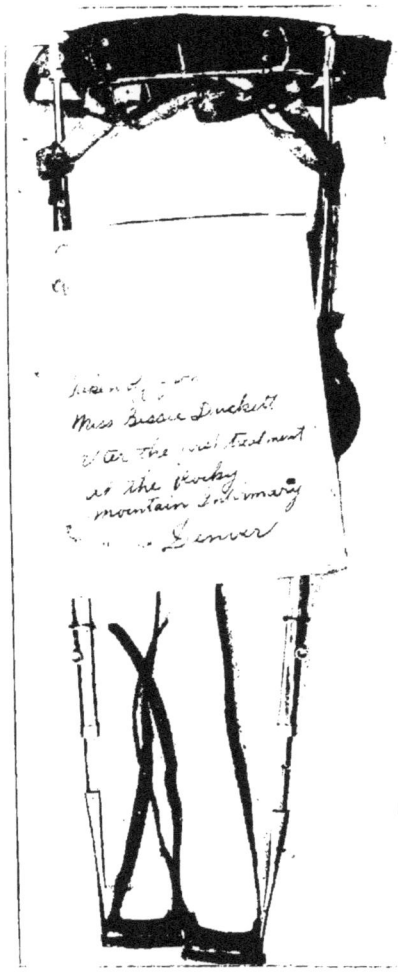

mained for over four months, growing continually worse
and gradually approaching a serious condition. Fortunately I was induced to bring the case to your attention.
Even after the first week's treatment I noticed marked
signs of improvement, and so continued to improve until
at the end of three months the injuries were completely
repaired. This cure I attribute entirely to your skill and
Osteopathy. Most gratefully yours."

"Denver, Colo., January 5th, 1897.
"Dr. W. L. Harlan, of The W. L. Harlan Rocky Mountain
Osteopathic Infirmary, Denver, Colo.:
"Dear Sir: Replying to your inquiry as to the results
of your first treatment of my wife for rheumatism, recently
given, I would say by way of preface, that for several years
past she has suffered greatly from that disease, and for the
past year or more has not been able to walk a step even
with crutches, or stand alone. We have tried all means
within our reach to cure her, but without success, the disease continually advancing until we had about despaired
of obtaining relief except that afforded by opiates, which
she had been compelled to use occasionally for several
months past, and daily for a month or more and for the
last week or two twice a day to control her pain. Your
first treatment, recently given, almost immediately produced marked improvement. Within a few hours she
brushed and did up her hair alone, a thing she had not been
able to do for almost two years. Her countenance, formerly pale and haggard, became fresh and rosy, and a general condition of comfort followed. She has since been
able to entirely dispense with the use of opiates, sleeps well
and restfully, and is so generally free from pain and so
greatly improved in all respects that we feel greatly en-
couraged and confidently hope for her speedy and complete
recovery. Hopefully awaiting results, I remain, yours
truly."

"Denver, Colo., August 7th, 1896.
"Dr. W. L. Harlan:
"Dear Sir: I desire to express my thanks to you for all
you have done in Mrs. W.'s case. She is better than she

has been for years. Your Osteopathic treatment has done more for her than all the doctors who have been treating her case for the past eleven years. Send all doubtful people to us and we will soon convince them of the benefits sure to accrue from Osteopathic treatment at your hands."

"Nevadaville, Colo., December 19th, 1896.
"Dr. W. L. Harlan:
"Dear Sir: It is with pleasure that I write to you to let you know that I appreciate what you have done for me. When I came to your infirmary, November 19th, 1 was suffering from stomach trouble in its worst form; also a severe pain in my chest, due to a strain from over-lifting. And now, after one month's treatment, I can say the stomach trouble has left me and 1 eat and sleep as heartily as ever, without feeling any signs of the trouble returning in the stomach. The pain in my breast is also gone, and I feel better than I have felt for years and have no reason to think the disease will return. Medicine could not do what you have done, as I have tried all the best doctors in the country, only to my sorrow."

"Dr. W. L. Harlan, City:
"Dear Sir: I want to say this much for Osteopathy. I have had catarrh for ten years or more and have doctored with specialists and have used patent medicines without receiving any relief whatever. I have had your treatment about six weeks and feel better, and my catarrh gives me less trouble than at any time since I first began to notice it, some ten years or more ago."

"Denver, Colo., November 17th, 1896.
"Dr. W. L. Harlan, City:
"My Dear Sir: In reply to your favor of the fifth inst., I am pleased to be able to say that your treatment relieved me of kidney trouble which had been with me for some time, and with which medicine did not seem able to cope with successfully."

"Denver, Colo.
"Dr. W. L. Harlan, City:
"Dear Sir: I have taken Osteopathic treatments from you, with the most wonderful results. Severe nervous trouble brought on by continued overwork and long residence in this high altitude was most effectually relieved by a short course of treatment. I shall be pleased to speak more fully to anyone desiring further information.
"Professor of Music."

"Denver, Colo., August 10th, 1896.
"Dr. W. L. Harlan:
"I find your treatment very beneficial for singers, especially where there is any contraction of the muscles of the throat, neck and chest, which condition frequently exists after an attack of 'grip.' The treatment restores the organs to their natural freedom.
"Most sincerely,   Music Teacher."

"Denver, Colo., January 2nd, 1897.
"Dr. W. L. Harlan, Denver, Colo.:
"Dear Sir: I commenced treatment at your infirmary the 23rd of August last for a stiff shoulder, pterygium on my eyes, and asthma. My arm and shoulder were in a very bad condition, and I was told there was great danger of permanently losing the use of my right arm. I suffered continually and at most times intense pain in my shoulder, elbow and wrist. For thirty days I have suffered but little pain, and am now able to use my arm quite as freely as ever. I am so greatly improved I will continue treatment, expecting to be fully restored to normal condition and to perfect health, which I thought a few weeks ago was impossible, particularly in respect to my arm and asthma. Hoping many others may profit by my experience, I would be glad to talk with anyone who may need the treatment. Most respectfully."

"Denver, Colo., 8-25-1896.
"Dr. W. L. Harlan, Denver, Colo.:
"Dear Sir: After receiving an almost fatal fall on October 2nd, 1895, I suffered with headache nearly every

day; also was troubled with my back, and after five weeks' treatment I am feeling as well as one could wish. The former troubles do not recur. I certainly believe that Osteopathy does what medicine taken into the stomach never can do. You may refer anyone to me and I shall be glad to state my belief from a logical standpoint more fully than I can here. Yours for the good of all."

"Denver, Colo., September 7th, 1896.

"Dear Sir: I took my daughter, four years old, to you to be treated for a dislocated hip of two years' standing. After two months' treatment she was greatly improved, and is still improving. I think the Osteopathic method of treating such conditions is far above any other.

"Yours respectfully."

"Senate Chamber, Denver, Colo., May 24th, 1897.

"Dr. W. L. Harlan, Rocky Mountain Infirmary of Osteopathy:

"Sir: Years ago I came to the conclusion, after spending considerable money in doctoring, that nervousness could not be treated successfully with medicine. Last winter, during the month of March, I tried a few of your treatments, and while they did not entirely cure me, they gave me more relief than anything I ever tried far in excess of my expectation, and I am satisfied that they have done a great deal for me and that Osteopathy occupies a very important and essential position in the medical world, and is a very important link in the chain that binds to this world, but more especially to health and happiness.

"Yours very respectfully,

"Senator 1st District."

"Arvada, Colo., January 4th, 1897.

"Having tried the Osteopathic treatment, I can truly say that I am satisfied that it has the merit claimed for it. My son was treated by you last autumn and was undoubtedly benefited. Yours sincerely."

"Denver, Colo., November 10th, 1896.

"In regard to Osteopathy and my case I can say it was miraculous the results I received through you. I went to you while I was in Denver. At that time I was suffering with severe pain in my back, which seemed to have come on in a peculiar manner and left me in a critical condition. Through a friend of the science I was persuaded to try Osteopathy, which at the time seemed to me a foolish idea. The idea of curing disease without medicine presented itself to me as an impossibility. But I thought to myself the world is advancing, and after you explained the theory of the treatment to me, I clearly conceived the idea that man is an engine and its divine construction is so arranged that if manipulated by a machinist who understands the human anatomy, and manipulated so that all parts will be adjusted and all forces pertaining to life are regulated, there is no reason in my mind why the system cannot be made to run harmoniously or be kept in a healthy condition, so long as all parts bear to their functions, and any of the numerous bones are not misplaced by accident, and the system not clogged or stupefied by drugs. To my mind nature is the great physician and the drug fraternity are hindrances to nature. Disease, as you explain it, is not a mystery, and is not to be dealt with through witchcraft and uncertain results of drugs, but must be dealt with by common sense and scientific methods. I can cheerfully recommend Osteopathy to any one suffering and would be glad to talk with any one who may want to know more of the treatment. Yours truly."

"Gunnison, Colo., July, 1897.

"My daughter Hazel four years ago had a severe attack of La Grippe, which left her left leg paralyzed, and by cutting off the nutrition to the leg caused it to stop growing and become perfectly cold. She has also a bad lateral curvature, beginning in the dorsal region and extending to the sacrum. She is also troubled with singing in the ears and at times very severe headaches. At the end of the first month's treatment we find the headaches and singing in ears entirely relieved, and the withered leg has assumed a natural heat by good circulation and can be brought down to the length of the good leg. Have removed an extra sole from the shoe and she walks better than with it.

"Yours truly."

"Boulder, Colo., November 10th, 1896.
"Dr. W. L. Harlan, Denver, Colo.:

"Dear Sir: I began treatment at your infirmary October 6th, 1896. At that time I could hardly walk, being troubled with pain in my back and hips. My circulation was also very much stagnated and caused me to have indigestion and a fainting sensation, and after eating I felt tired and sluggish. My feet also were cold continuously, so that I would have to sit close to the fire most of the time in order to keep from being chilly. I also was suffering from chronic diarrhœa, which I had doctored for ten years, and after the first treatment I was entirely free from all annoying symptoms, and went with my husband and friends to view a band parade and contest, a thing which I could not possibly have done before taking the first treatment; but I kept on with the treatment for a month in order to relieve my system of all impurities and build up my general health, so nature could ward off the disease and keep it from returning. When presenting myself at your infirmary you recommended that I discontinue the use of all drugs from any source whatever, or treatment with any other physician, as you wanted your treatment to have full effect. You told me that drugs only served to clog and stupefy the system. I followed your advice, and now after one month's treatment I can say I am feeling so much better, the pains and former troubles have disappeared, my friends even remark how much better I look. To my mind Osteopathy is the greatest discovery of the age, and I can cheerfully recommend the treatment to any one suffering from my trouble, and will be glad to talk with any one needing the treatment.

"Very truly."

"Gunnison, Colo., July 6th, 1897.
"Dr. W. L. Harlan:

"I have suffered with very severe headaches and backaches for a number of years, caused by a strain at sacrolumbar articulation. Was also bloated so I could scarcely button my skirts around me. Eyes were sore and weak and had not been able to get a good night's rest for years. Also have very severe nervous spells lasting sometimes for three or four days. After one month's treatment I find a very decided improvement in every way. Backache and headache entirely gone; nervousness and bloating improving. Eyes are very much improved. Yours truly."

"Denver, Colo., May 19th, 1897.
"Dr. W. L. Harlan:
"Our boy met with an accident five months ago. In falling he dislocated his hip. I took him to our family physician. He told me we would have to put him in a splint. Keep him there for two years. Went to another physician. Said the same. Then took him to one of Denver's best surgeons. He said that even a splint could not cure him, that he would be a cripple anyway. Went to you as a last resort. You told me you thought you could cure him in three months' treatment. It is two months since we began taking the treatment. Thanks to you and Osteopathy. Our boy is cured completely. Will be pleased to talk with anybody on the treatment. Yours truly."

"Denver, Colo., June 4th, 1897.
"Dr. W. L. Harlan:
"Under your treatment I have been relieved of constipation; several sore spots in the region of the spine have been cured; my circulation has been helped, and my general health is greatly improved. Believing that your system has much merit, I take pleasure in recommending it."

"Manitou, Colo., November 6th, 1897.
"Dr. W. L. Harlan, Denver, Colo.:
"Dear Sir: I can say that I have entirely recovered from the attack of rheumatism which I had some time ago and am satisfied that your treatments were the cause of my speedy recovery as I used no other remedy. You can rest assured that either myself or husband will be glad to, and will speak a good word for you or any other Osteopath who graduates from Dr. Still's school, whenever an oppor-

tunity presents itself. You can refer any one you desire to me and I will gladly tell them what I know about Osteopathy. Wishing you success in your new location, I am, "Respectfully yours."

"Denver, Colo., August 15th, 1896.

"My case is rather an odd one. I came here from Brooklyn for lung trouble and was here three months without any improvement to speak of. Hearing from many people of your success, I made an effort to see you. The result is, that my cough was stopped in three treatments, which I thought a few weeks ago was incurable. I am now as strong and well as ever, after nearly three weeks' treatment, but will continue to take them until I know I am perfectly cured, which I have reason to believe will be in about three months. I recommend this treatment to anyone suffering from lung trouble, for it has relieved me since taking it, and I expect it will cure me of what I thought impossible a short time ago, and this, in the face of the fact that my case was pronounced incurable by doctors both here and in Brooklyn, N. Y. Yours sincerely."

"Gunnison, Colo., July 1st, 1897.

"I came for Osteopathic treatment one month ago, suffering from heart trouble. Left shoulder was very lame, and had very severe headaches and catarrh. The eleventh rib on the left side was jammed up under the tenth cartilage, causing a dull pain at the ensiform appendix. After one month's treatment I find a general improvement in all symptoms."

"Denver, Colo., June 23rd, 1897.

"While at college I had a fall which caused spinal curvature. Have had occasional attacks of spinal meningitis and nervous prostration, and have been treated by the most able physicians of New York, Boston, Chicago, St. Louis and Detroit, besides have taken electrical and magnetic treatments, massage, mineral baths, etc., but nothing gave me permanent relief. After six weeks' treatment from you, I have been entirely relieved of the curvature of the spine and pressure of the blood against the brain, and the somewhat distorted condition of the body has been removed and it is regaining its symmetry. It is absurd to take medicine when an Osteopath is needed to set a misplaced bone. Osteopathy is one of those indescribable modes of treatment and an Osteopath can do for a patient what a physician or surgeon can not do. I would be glad to talk with any one suffering from curvature of the spine or nervous prostration.

"Elocutionist, Boston, Mass."

"Denver, Colo., August 17th, 1897.
"I have had but one month's treatment for bronchitis by you. Can say I am much better; think Osteopathy is all that its followers claim. Anyone wishing to see me can find me at the Brown Palace Hotel. Doctored for years with other doctors who did me no good; this treatment relieved my cough right away. Yours truly."

"Senate Chamber, Denver, Colo., April 11th, 1897.
"W. L. Harlan, Doctor of Osteopathy, Kittredge Building, Denver:
"Dear Sir: During the month of March I was taken down with La Grippe in its worst form, and my friends telephoned for my wife, hardly thinking I would live long enough for her to get here; after sending for her you were called in, and in fifteen minutes removed the most violent headache that a human being ever suffered with and in half an hour removed every pain from my body. A pleasant sweat was produced that continued for twenty-four hours, so that I had to change my underclothing twice in that time. All this was done without the use of one drop of medicine of any kind. In five days I was able to resume my duty in the Senate and attend its all-night and day sessions without any more difficulty than if I had not been sick. This remarkable cure and treatment calls to my mind that I suffered with La Grippe six years ago and was treated by one of the old school doctors. I suffered the severe effects of that disease for over six months and did not feel as good at the end of one year as I do at this time. In the case of Osteopathy it required two visits to produce

16

present results. In the case of the old school doctor it was two to three visits daily for weeks, and large quantities of drugs to be swallowed. Yours very truly,

"Senator 12th District, Silver Plume, Colo."

"Detta, Colo., June 21st, 1897.

"Dr. W. L. Harlan, Rocky Mountain Infirmary Association of Osteopathy, Denver, Colo.:

"Dear Sir: It gives me pleasure to testify to my appreciation of the benefits of Osteopathy received through treatments last year at your hands. While I was unable to continue under your care as long as I desired, still, during the nearly two months you treated me I never received a treatment without experiencing direct benefit. I would heartily recommend your science to all sufferers. Yours."

The following are a few of the diseases which have been successfully treated by Osteopathy: Rheumatism, Lumbago, Gout, Backache, Stiff Neck, Kidney Diseases, Atrophy of the Limbs, Eczema, Diseases of the Bladder, Retention of Urine, Paralysis, Impotence, Curvature of the Spine, Stiff Joints, Stomach and Bowel Troubles, Rectal Troubles, Chronic Diarrhœa, Female Diseases, Milk Leg, Constipation, Peritonitis, Dysentery, Piles, Appendicitis, Torpid Liver, Gall Stones, Dyspepsia, Results of La Grippe, Nervousness, Nervous Prostration, Restlessness, Insomnia, Hysteria, Convulsions, Varicose Veins, Cold Feet, Asthma, Bronchitis, Hay Fever, Polypus of the Nose, Catarrh, Deafness, Diseases of the Throat, Sore Throat, Enlarged Tonsils, Goiter, Neuralgia, Sick or Nervous Headache, Epilepsy, Heart Disease, Dizziness, Granulated Eyelids, Dripping Eyes, Lung Trouble, Difficult Breathing, Pterygium, Lameness of all Parts.

www.ingramcontent.com/pod-product-compliance
Lightning Source LLC
Chambersburg PA
CBHW021524210326
41599CB00012B/1368